Growing Bread on Trees

The Case for Perennial Agriculture

Dr. Elaine M. Solowey

Front cover photo credits:

Oak Forest—Geoffrey Hobson
Pine Forest—Paula Keys
Pomegranates—Guy Eisner

ISBN 978-0-9833352-0-7

Substance Editor – David Schutt, engineer and horticulturist

The Thistle Syndicate is a group of writers and researchers who are dedicated to bringing specific skills and information to the public in a series of low cost publications.

Biblio Books

This book is dedicated to Ed and Vivian whose generosity makes my work possible.

And to Serdar Afacan, Ana Schwartz and Hannah Medalia who work with me every day.

Table of Contents

Introduction
The Case for Perennial Agriculture

Proposed in this book is a form of agriculture just as productive as modern agricultural systems but that rests more lightly and kindly on the suffering earth.

In this system no wasted nutrients pollute the water we drink, no topsoil is blown away by the wind and crop- protective chemical sprays do not contaminate the food we eat.

This system of agriculture draws carbon from the overloaded carbon cycle and stores it usefully and safely, stabilizes climate and greatly reduces the need for artificially fixed nitrogen.

In short, it is a way to make almost all the food products we need from edible oil to carbohydrates, from salad, to sweets to nuts.

We can also produce animal feed, rope, fabric, medicine, spices, teas, fuel and building material.

And while we are doing this we can stop erosion and greatly slow global climate change.

The only obstacles to using this agricultural system to heal the earth is our ignorance of what perennial crops can do for both us and the land we work—and our attachment to a kind of agriculture that was developed in the stone age and has changed since then only in scale and intensity.

The Weakness of the Current System

Modern agriculture is not sustainable because it destroys topsoil and uses more energy than it produces in the form of food.

This is a fact, not a supposition and if the truth about modern agriculture is not faced squarely everyone fed from this system is going to suffer.

Other factors besides erosion and the negative energy equations of modern food production exacerbate the agricultural situation and it is only a question of which set of circumstances will burst the bubble first.

Some serious thinkers believe the increasing cost and scarcity of oil will lead to crisis.

The following troubling quote is from a book called *The Long Emergency* in which author James Howard Kunstler envisions a bleak future brought on by the shortage of oil.

He assumes rightly that modern agriculture is exceedingly dependent on non- renewable fossil fuels for transport, energy to run irrigation systems and farm machinery and for the manufacture of fertilizers and pesticides. He is also extremely critical of the way modern agriculture is being done:

> *"There is a reason why farming is called agriculture. The culture part stands for the body of knowledge, skill, principles and methodology acquired over thousands of years. Most of that knowledge has been jettisoned in a rush to turn farms into something like automated factories. In fact the current system is explicitly called "factory farming" by those who run it. The technology of factory farming promotes the expansion of farms by orders of magnitude above what had been the upper limit for traditional non-industrial farms. Increasingly farming has changed from being organized on a family and community basis to being corporate and national, even global with few benefits for the localities where it takes places and with devastating effects on local ecologies and social relations. The diminishing returns of technology in farming have been especially vicious. Few other human activities*

demand so much respect for natural systems and the abuse of natural systems has been monumental under the regime of industrial farming. The genetic modification of monoculture crops is only the latest (and possibly the final) technological insult among many previous ones and comes at the climax of the of the industrial blowout. Diminishing returns are nature's way of biting back. The "winners" in recent decades have been the corporations that could enjoy the economies of scale conferred by gigantism. The "losers" can be summarized generally as the future and its inhabitants. They stand not only to lose future wealth but their civilization." James Howard Kunstler.

Jared Diamond is another author who believes that agriculture must be changed radically. His recent book **Collapse** details the rise and fall of several societies as diverse and separated by geography as Mesa Verde in the American Southwest, Easter Island in the middle of the Pacific Ocean and the Viking settlements of Greenland. In each case natural systems were abused, trees were cut down and agriculture endeavors extended far beyond the point of sustainable land use. Strained to the tipping point, droughts, bad weather, and social unrest brought these societies down. The greatest factor contributing to the failure of these cultures was the collapse of the local ecology and agricultural systems that sustained them.

Jared Diamond obviously believes that ecological and social problems are just as dangerous for our own unsustainable systems as they were for the now extinct Greenlanders.

The result of such a failure now would be truly catastrophic. The percentage of people who actually know something about producing food in the developed world is as low as it has ever been in human history. Left to fend for themselves, 99% percent of the current population would probably starve to death in a few weeks since so few of them would be able to grow, gather or catch anything to eat.

Whether survival would be possible even for the agriculturally skilled few- as society melted down around them- is debatable.

Both authors predict failure for modern conventional agriculture. They are not alone in this opinion. Since modern agricultural operations are so disrespectful and wasteful of the resource base that supports them more than a few scientists have come to similar conclusions.

Without cheap oil factory farming, global trade and many other systems we take for granted will not be possible. And without these systems collapse is inevitable, therefore the scientists opine, what we call civilization will not survive the deadly convergences of the 21st century.

Certainly without cheap oil there will be no commercial aviation, no monster wheat combines, no rush hour, no private motor vehicles and probably no skyscrapers. Lettuce will not be trucked 2,000 miles to the dinner plate and cheap appliances will not be imported from China

But civilization existed long before these things were invented and agriculture, which made civilization possible, is even older than that. I personally do not believe that human beings are helpless in the undertow of approaching problems, like so lemmings being swept out to sea. After all, it is humankind who invented economics, globalism, factory farming, genetic modification, spaceflight and a host of other wise and unwise methods and endeavors.

If these things have ceased to serve our purposes then they can be changed—or if necessary, **uninvented** by exercising restraint and not using whatever piece of destructive cleverness is causing the problem. The idea that progress is an irresistible force pushing us toward a disaster from which we cannot turn aside is an adolescent and unworthy notion.

The tendency of certain people including scientists, politicians and philosophers to parade about with end- of- the-world signs pinned to their clothing will not help humanity through the coming challenges of climate change, fossil fuel depletion, political unrest, periodic epidemics, droughts and famines. Sound, clean, wise and respectful systems of agriculture- if implemented immediately- will see most of us through the crisis and will heal and renew a good piece of the planet as well.

Agriculture does not have to be inherently destructive.

In fact there is one sort of agriculture that is almost always regenerative, positive and renewing and that is the practice and culture of growing trees.

Why then is this not the agriculture that we practice?

The tree in the landscape

Erosion is the enemy of both agriculture and civilization according to J. Russel Smith who made a series of expeditions to the Mediterranean, Far East and the Middle East to study various patterns of land use. He was appalled by the vast stretches of destroyed and depleted land he saw.

> *"Forest—field—plow—desert... that is the cycle of the hills under most plow agricultures. China has a deadly expanse of it (destroyed land) but so has Syria, Greece, Italy Guatemala and the United States. Indeed we American though new upon the land are destroying soil by field wash faster than any people that ever lived-*

ancient or modern, savage, civilized or barbarian. We have the machines to help us destroy as well as create. The merciless and unthinking way we tear up the earth suggests that our chief objective may be to make an end of it.

How does it happen that hill lands have been so frightfully destroyed by agriculture? The answer is simple. Man has carried to the hills the agriculture of the flat plain. In hilly places man has planted crops that need the plow and when a plow does its work on lands at an angle instead of flat lands we may look for trouble when rain falls." J. Russel Smith. *From Tree Crops, A Permanent Agriculture*

It should be added at this point that when crops need the plow on flat lands we should also look for trouble when the wind blows. Second only to water erosion is the relentless erosion of arable lands by the wind which carries away millions of tons of soil each plowing season.

Between the water erosion on hilly and sloping lands and wind erosion of flat plains the fertile topsoil of the world is either drifting away or being washed away.

The population of the world is as large as it has ever been in human history and all of humanity needs food. But depending for our food on a system that systematically ruins the land it uses simply means that sooner or later this kind of food production will simply not be possible.

In my opinion erosion of the topsoil is as big a terror as abrupt climate change, a vast underlying problem not as visible as the coming shortage of oil but a far greater threat to humankind.

We practice an agriculture that kills the soil, depletes the soil, makes the soil disappear.

And we are utterly dependent for our food on a resource that we are destroying with almost every kind of crop we plant.

The megacrops of the 20th century are all weak competitors and have to be planted in environments where other plants have been eliminated. Cereals for instance are grown on bare fields with every weed sprout sprayed or harrowed out of existence. Rice is cultivated in flooded, intensively weeded methane producing paddies, soybean and rape seed are grown in fields made abiotic by chemical application and corn is grown in huge monocultural blocks as big as small countries where no other living thing is allowed to survive.

A terrible ecological price is paid for this kind of cultivation. Communities wither, wildlife disappears and the land becomes barren. Unfortunately almost every modern crop is cultivated in this manner.

Every kind of crop except the tree crops. Tree crops, besides their bounty of fruits, nuts, fibers or pods, are a blessing to the naked, wounded earth.

Though there are many ecologically unfriendly and unsustainable orchards, tree cultivation is not inherently damaging in itself. Annual plowing is not needed. Agricultural run off is not a problem in a well tended orchard. Trees do well with long cycle nutrition so trees can be composted and mulched instead or fertilized with chemical preparations. Diverse orchards can be kept pest free with sensible IPM strategies.

Just about every damaging factor in modern agriculture is absent from arboreal cultivation. And the ecological benefits of the trees are profound and numerous.

Trees make food and air

Trees turn sunlight into stored food and fuel by photosynthesis, making sugars, carbohydrates, cellulose and many other substances that nourish insects, wildlife, domestic animals and human beings. Useful non edible tree products are too numerous to name. Trees absorb carbon of which there is currently too much in the atmosphere and produce oxygen on which all higher animals depend.

Trees manufacture topsoil

Trees are rooted deeply in the earth, like living pumps, breaking up rocks in the subsoil layers and releasing minerals needed to make topsoil. The fine root hairs absorb these minerals and the tree brings up this food and stores it in the form of fruit, seeds and leaves. The fruit and seeds become food for animals, birds and man but the leaves fall to the ground and rot, releasing the stored minerals into the soil for other plants to use. The roots of the tree stabilize the soil and become the refuge for soil-making microorganisms and insects.

Trees stop erosion and store water

This zone of biological richness full of roots that penetrate the ground and make channels and pathways for water infiltration then becomes an area which can absorb and hold run off allowing the tree to grow and develop and the soil creating activities of all the associated organisms to continue. With the subsoil broken up by the powerful rooting system excess water can flow down in the subsoil and eventually to the aquifers.

Trees are magnets for rain

Trees draw up water from the ground and make it available for the next cycle of precipitation by breathing out water vapor into the air. The famous naturalist who described the Amazon basin as the lungs of the world was not speaking poetically. He was being accurate. A mature deciduous tree may transpire five hundred liters of water from its approximately six acres of surface area on a warm summer's day. The cool air under the trees covering a forested area, drawn upward as rain clouds flow over, often encourages the formation and fall of raindrops. When the rain falls the tree will intercept it, absorbing water through the leaves, shedding the raindrops downward to pool around the trunk and roots.

Trees are home for beneficial creatures

Trees provide a living structure to inhabit for innumerable plants and animals. Their leaves and products feed many more creatures and their roots and root structure shelter even more. Biologists investigating the rain forest dramatically revised the number of estimated species of birds and insect they believed to exist after discovering thousands of new plants and new creatures in the canopies of old growth forests. Burrowing mammals and insects, benign fungi, nematodes earthworms and dozens tiny microorganisms including soil building and nitrogen fixing bacteria find homes in the branching, permeable and complex underground structures of trees.

Trees mitigate and stabilize the climate

The air and the earth are several degrees cooler under a tree in hot climates and several degrees warmer in a cold one. As well as acting as the lungs of the world, trees are the world's air conditioners. They slow and divert air currents so that dust particles drop out, cool and add moisture to the scorching wind of the desert. Forested slopes are prime areas for fog condensation. A north wind, tamed by trees, is more likely to murmur than roar. What trees do around our dwellings they also do for the world protecting, shading, stabilizing and cooling.

Deforestation, something of a throw-away line when global climate change is discussed is most likely one of its primary causes.

With all the benefits of arboreal agriculture why then is plowed agriculture the most prevalent form of cultivation?

And with the loss of forest cover one of the major contributing factors to climate change—- why are no alternatives being seriously discussed?

Chapter 1
Forests and Climate

Deforestation rapidly and completely changes both the climate and the geography of the areas where forests of trees have been cut down. Deforestation is the primary cause of floods, mudslides, land slips, avalanches and other similar disasters. It is also one of the factors that affect the amount and pattern of local precipitation.

Deforestation was the main cause of the decline of the Byzantine era city of Ephesos, now a ruin in modern-day Turkey. Trees were cut down on the surrounding hillsides for shipbuilding and other uses causing the erosion which first made the harbor shallow, then turned the harbor area into a malarial swamp, and which ultimately left Ephesos sitting six kilometers inland with no access to the sea at all.

It has been known for many hundreds of years that deforestation has dire consequences. As early as 1215 Louis the VI of France promulgated an ordinance called "the decree of water and forests" which promoted conservatory measures along river banks and forested watersheds. He was of the opinion that deforesting these areas led to floods and that as divine retribution for allowing these disasters to happen, God would stop sending the rain.

Christopher Columbus was also of the opinion that forests promote precipitation. He commented on the decline of rainfall in the Azores and Canary Islands after Spanish settlers felled much of the natural forest cover.

It has also been known for centuries that the health and extent of forests influences the rivers and streams that flow from watersheds. An Italian engineer by the name of Lombardini measured the amount of silt and soil carried to the sea by the Po River and found that 25 million cubic meters of soil had been washed away after the clear cutting of the nearby mountain forests.

Cutting trees on watersheds in Italy, Austria, and France was forbidden as early as the sixteenth century as it became apparent that deforestation caused erosion and the over silting of rivers. Germany soon followed with anti-clearing laws after publications appeared on the subject.

Reforestation projects, as a means of controlling seasonal flooding, began in Japan in 1683 and the practice of reforestation was enforced by imperial decree.

In the 1700s and 1800s the sand dunes on the west coast of France were reforested. These dunes had originally been covered with trees. When the forests were cut down the dunes began to shift, covering up agricultural land and threatening villages on the coast. Small scale attempts at reforestation were made by private parties whose property was threatened. Then the Commission of Dunes was created in 1799 to carry on the work systematically and with the help of the government. In this manner, 200,000 acres of shifting sands were fixed and reforested. The nearby Landes marshes, almost as problematic and considered a breeding ground for malaria, were reforested between 1837 and 1892 by planting 1,750,000 acres of maritime pine.

In 1837 in Denmark a study of the deteriorating climate of Jutland attributed the decline to the felling of most of Jutlands natural forest. This had exposed the peninsula to strong sea winds that blew the coastal sands inland and buried large areas of fertile soil.

A treatise on forest influences was published in 1853 by French scientist Becquerel. In this work he asks serious and pointed questions about the influences of forests on the surrounding areas:

1. What is the part that forests play as shelter against the winds or as a means of retarding the evaporation of rain water?
2. What influences do the forests exert through the absorption of their roots or the evaporation from their leaves in modifying the hydrometrical conditions of the surrounding atmosphere?
3. How do they modify the temperatures of a country?
4. Do the forests exercise an influence upon the amount of water falling and upon the distribution of rains through the year, as well as, upon the regulation of running waters and springs?
5. In what manner do they intervene in the preservation of mountains and slopes?
6. What is the nature of the influence that they may be able to exercise on public health?

Becquerel did not answer these questions to his own satisfaction in a lifetime of labor though he did not think reforestation could revive vanished springs and streams.

A contemporary of Becquerel by the name of Von Wex, studying the Danube basin of Central Europe, came to the opposite conclusion. He recorded reductions in the level of water in wells and streams when lands were deforested that were partially restored in areas that had been replanted.

In the United States the problems of deforestation were recognized as early as 1739 when a series of laws were passed against cutting wood and grazing animals too close to the Massachusetts shore.

Denuding the areas near the coast had allowed the sands to blow inward and bury fertile fields.

In 1799 Noah Webster presented a paper to the Connecticut Academy of Arts and Sciences on a perceived change in winter temperatures he attributed to deforestation. He said:

> *"From a careful comparison of these facts it appears that weather in modern winters in the United States is more inconsistent than when the earth was covered with woods at the first settlement of the Europeans in the country, that the warm weather of autumn extends further into winter and spring encroaches upon summer; that the wind being more variable, snow is less permanent and perhaps the same remark may be applicable to the ice of the rivers. These effects seem to result necessarily from the greater quantity of heat accumulated in the earth in summer since the ground has been cleared of wood and exposed to the rays of the sun and to the greater depth of frost in the earth in the winter by the exposure of its surface to the cold atmosphere."*

In short Noah Webster is bemoaning the loss of the forested woods as a mitigating factor on the weather. Without the forest, the weather is both hotter and colder and also windier and drier. His paper was one of the first published references to the subject in the United States.

Webster's concerns were echoed by writer W.C. Bryant in 1858: "**Streams are drying up and from the same cause, the destruction of our forests, our summers are growing drier and our winters colder.**"

In 1877 Dr. R.B Hough prepared a report for the US government on the subject of forestry that included a lengthy section on the connection between forests and climate complete with compiled evidence to show the beneficial effects of forest cover and the deterioration of the local climate when the trees were cut down. Senators and representatives who read the report were impressed and interested enough to try to pass a bill in the US Congress for the preservation of the forests of the national domain adjacent to the sources of navigable waters and other streams in the US. The aim of this legislation was to keep the rivers from drying up.

The bill did not pass, but in 1882 a similar bill passed in Massachusetts. California was not far behind. California passed a bill in 1892 which reserved 17,500,000 acres of forested watershed in the south of the state. These were the first laws passed in an enthusiastic initial effort to preserve some of North America's vast forests.

In 1902, the Forest Reserve Manual gave these reasons to maintain forests on watersheds:

1. To furnish timber
2. To regulate the flow of water. This the forests do:
    a. By shading the ground and snow and affording protection against the melting and drying action of the sun.
    b. By acting as windbreaks and this protecting the ground and snow against the drying action of the sun and wind.
    c. By protecting the earth from washing away and thus maintaining a storage layer into which rain and snow soak. This storage layer conserves the moisture for dry seasons when rain and snow are wanting.
    d. By keeping soil more pervious so that the water soaks in more readily and more water is thereby prevented from running off in time of rain and when snow is melting.

In 1911 the Weeks Law provided for the protection of watershed of navigable streams and the acquisition of lands for conserving their navigability. The eastern United States was convinced of the value of forests to the general health of riparian areas.

After the man-made disaster of the 1930's Dust Bowl, the Western States were much more likely to listen to advice about soil stabilization, erosion and flood control. Many deforested areas were replanted with trees and huge areas, previously gone to dust, were replanted in native grasses. Unfortunately, there were and are still too many people who believe that God put trees on the earth to be cleared off building sites and processed into paper and lumber.

An example of this attitude can be found in the hills around modern Athens. Every summer fires burn on these mountains around the city. This is not mysterious, since, by their laws, land which has been "burned over" becomes eligible for rezoning and building, who needs these useless trees which are standing in the way of "development?" The predictable result of this annual burning is a significant decline in precipitation on the seaward side of the mountain range and a serious erosion problem as tons of topsoil are washed into the sea. The consequences in 2007, during an especially dry year were horrific, as fires set to "burn over" real estate raced out of control, charring thousands of acres and killing dozens of hapless citizens all over Greece.

In the case of the fires around Athens, we are only speaking about the further ruin of sparsely forested areas in Greece, a country more or less deforested in ancient times. However, the destruction of the rain forests, the lungs of the world, for temporary farmland, is widespread, ongoing, and apparently unstoppable. This destruction is a critical loss.

To fully understand the outcome of the loss of the rain forests, it is necessary to consider the interaction of one living tree in the landscape on its contiguous environment and the many beneficial effects of that interaction. Because of this one tree: the ground is made permeable, moisture is captured and allowed to seep into the earth, the sun's energy is absorbed and turned into food, water vapor is released into the air, carbon is stored and oxygen is manufactured, the earth is shaded and cooled.

Now, imagine this tree has been cut down. It its place is a barren spot which has been left by cutting the tree. No food is made there. No oxygen or water is released into the air there. No carbon is stored there. The very earth becomes hardened and impervious to the rain. The hot sunlight striking this naked place is being reflected back into the air, heating the air up. Now multiply the effects of the loss of one tree one million, five million, ten million times over. This is a tragedy. With all that is known about the effects of deforestation, from flooding to loss of rainfall, silted rivers, harbors made too shallow to receive cargo ships, it should be apparent to all that deforestation has profound and long lasting deleterious effects.

While it is argued by some that climate, geography and water storage capacity are only modified on a localized basis by stripping the nearby land of protective trees, many hundreds of locally modified climatic zones quickly add up to vast stretches of denuded and depleted land. Collectively, these damaged landscapes have become so vast that the "redline" of the critical amount of surface area that must be maintained in a forested state for sake of the good health of the global environment is in danger of being surpassed.

The elements of climate that are most significantly affected by the presence or the absence of trees are water retention and precipitation, solar radiation, temperatures and wind. In other words, what is most affected is the amount of rainfall, the ability of the land to store water, the ability of the aquifer to renew itself, how much sunlight is converted to biomass and how much is reflected back into the atmosphere to heat up the air. Also affected is the strength, direction and temperatures of the winds and even the shape of the landscape itself.

With all these critical factors changed so profoundly by deforestation, it is hard to understand why the effects of deforestation on the global climate are not taken more seriously by everyone.

I do not dispute the fact that our penchant for fouling our only planet by burning fossil fuels is causing damage. I am also convinced that the fact that there are fewer trees to absorb carbon and fewer trees to produce oxygen and fewer trees to encourage rain and fewer trees to decrease the amount of CO2 in the atmosphere is just as critical a factor in our environmental damage as is our fossil fuel consumption.

We know by historical example that changes in climate, particularly decline in rainfall and the resulting reduced flow of springs and stream are the direct results of large scale deforestation. The silting up of harbors and damage to property by shifting sands are a direct result of deforestation. Damage to life and property by flooding, torrents and avalanches are common consequences of the cutting down of forests, especially on sloping land.

Sadly, many of the ways that trees mitigate the climate are only now beginning to be understood. Australian scientists have recently discovered, for instance, that upland rainforests harvest vast amounts of water directly from clouds in addition to the water that falls on them as rain. In high, wet, tropical areas above 900 meters, low clouds, mists and fog constantly condense on the trees and the drops of water are soaked up by the ground when they collect and run down to it. This does not happen when areas are cleared of their forest cover. Note that this cloud water is *in addition* to the water that falls as rain. Because these upland forests transpire very little they contribute a disproportionately large volume of water to their catchment area. This contribution is greatly reduced when the forest is cleared and the water is then allowed to escape by running off.

Experiments carried out in North Queensland and reported by the CSIRO placed rain gauges, troughs and collar gauges around and among the trees, measuring direct rainfall, condensation on the trees, and stemflow, or the water than runs down the tree trunks. The amount of water "stripped" from clouds, fog and mist by the forest was far greater than expected.

Because of these experiments it is only now known that if the cloud banks, which currently contribute their vast quantities of water to the forest via cloud stripping by the trees, should rise in altitude due to global climate change or should be lost because of deforestation, there will be a major loss of water to the catchments and all the communities downstream.

Another little known mitigating factor only recently understood is the protective influence of the mangrove swamps and estuaries on fragile coastlines. These partially inundated areas where the water may be as shallow as a few inches or several meters deep according to the tides, are a tough natural buffer that protect the coastlines from wave action and weather events. Areas where the mangrove trees have been cleared for shrimp farms and fisheries are particularly vulnerable to erosion, storm surges and flooding.

The 2008 Burmese cyclone, a catastrophe that reportedly killed over 100,000 people was apparently exacerbated by the deforestation of the tidal mangrove forests. Without the deep strong fibrous roots of the trees to prevent erosion the coastline was literally eaten away allowing the surging tides to roll over the low lying areas and penetrate far inland. Far from being useless and unproductive area, the mangrove forests were a necessary barrier, a protective zone that shielded the land from the action of a periodically aggressive sea.

What has not yet been grasped by much of the world community is that there is a tipping point, an amount of tree cover which must be maintained globally lest the climate change too radically and too irrevocably to recover. A lovely fable was written in the year 1953 by Spanish author Jean Giorno. It was called The Man Who Planted Trees. In this deceptively simple story a solitary shepherd whose wife and son had died roams a desolate land with his flock. However, he does not accept the abandonment and ruin around him. Every day he rescues seeds and acorns from the wasted landscape sorting them at night so that only the perfect, healthy ones remain. On the following day he plants one hundred trees. He does this day after day, year after year until the barren land is again a beautiful forest and the springs and rivers are restored and people come back to live in the land again.

This was such a compelling tale that many people thought the shepherd and the forest were real and tried to contribute money to the project or visit the restored area. This particular story is fiction but forests and woodlands have been restored all over the world with very encouraging results.

What the fictional fable describes is truly possible. The barren land can be made green. The air can be made clean and the water cycle renewed. With great effort a wounded landscape can be healed. Yet, consider how much better it would be if we simply paid attention to the hard lessons of the past 2,000 years from Ephesos to the Amazon and acknowledged the fact that we need trees, not just for wood and forest products but as great living stands of treed forests to the keep our world healthy, our air breathable and our global climate stable.

Chapter 2
Carbon Sequestration

Carbon is one of the several elements which cycle through the natural world. The carbon cycle is considered the first of the biogeochemical cycles. Carbon is the basic building block of life. It is the cycling of carbon from the atmosphere through plants and algae, to animals and microorganisms and then back to the atmosphere which keeps the atmosphere and climate in balance.

During photosynthesis plants combine carbon dioxide (CO_2) from the air and hydrogen from water to make carbohydrates ($C_xH_xO_x$). Some of these carbohydrates are used directly by the plant for its energy. Others are stored in plant tissues in the form of sugars, starches, and oils. When plants are eaten by animals, the plant tissues, with the stored carbohydrates, are broken down by the animal's digestive system. The plant's stored carbohydrates and other nutrients are absorbed into the animal's nutritional system. The carbohydrates are "burned" as fuel for the animal's body in the animal's cells and end up in the form of carbon dioxide once again. As animals exhale, the carbon dioxide is released into the air where it can be reused by being reabsorbed by plants.

If this was all that was happening in the natural world, the carbon cycle would be relatively stable and balanced. However, carbon also gets back into the atmosphere when plant and animal matter are burned. Fossil fuels, such as oil, coal, and natural gas, are made of ancient, compressed, and concentrated plant tissue, the stored food made by years upon years of the energy of ancient sunlight and the carbon removed from the atmosphere millions of years ago.

In the past century the carbon dioxide content of the earth's atmosphere increased apparently by about 25%. It is continuing to rise 0.4 percent per year. If the present rate of increase continues, the pre-industrial atmospheric content of CO_2 will be doubled by about 2150.

Since increased CO_2 in the atmosphere is one of the factors believed to promote global climate change, especially the so-called greenhouse effect, this increase is seen to be very threatening to climatic stability. Indeed, some scientists have begun to relate to CO_2 as if it is a deadly pollutant (which it is not) and some economists foresee carbon credits as a new form of currency (which they should not be).

Still, increased carbon dioxide in the air is a factor in our changing world. It has been estimated by the members of the IPCC (the United Nation's International Panel on Climate Change) that average mean temperatures near the Earth's surface will increase between 2.8 and 5.2 degrees centigrade with a doubling of present carbon dioxide levels. However, it seems scientists and commentators cannot agree on whether this is a "cause" or an "effect," whether increased carbon dioxide will lead to warming or whether a warming globe simply releases more CO_2 into the atmosphere from warmer oceans and melting permafrost. Indeed, there are many serious scientists who believe the opposite of global warming is currently happening. Recorded temperatures are actually falling, these scientists show, because of a dimming solar cycle and the globe is measured to be cooling despite the increased CO_2 in the air.

Whether the globe is cooling after warming for the last few decades, or warming up after the Little Ice

Age (ca CE 1300 – 1850), is almost a moot point. Increasing tree cover stabilizes climate no matter where we are on the climatic roller coaster and should be something of a first line of defense. It is amazing that "geo-engineers" can speak about putting reflective particles in the atmosphere or painting roofs and roads white as climate mitigating strategies while forgetting the only strategy for stabilizing climate that has been shown to be both effective and benign.

Yet, whether carbon is the cart or the horse is also unimportant. Each year the burning of fossil fuels sends approximately twenty billion tons of carbon dioxide into the atmosphere, compared to estimated emissions of only seventy two million tons per year a century ago. Since WWII the use of fossil fuels has doubled roughly every ten years. The present destruction of tropical forests and old growth areas is probably releasing another five billion tons of carbon dioxide annually. While much of the CO_2 is absorbed by oceans and other carbon reservoirs, about eleven billion tons remain in the atmosphere, already causing changes in the biosphere and serious risks to human health.

While adjustment to these increased levels in some ecosystems is occurring, it would be wise to halt the CO_2 enrichment of the atmosphere as soon as possible since we do not know what will be the outcome of a continuing rise. We also do not know what level might be a tipping point for unforeseen changes. Furthermore, we do know that polluting the air with all the other substances that are released with the burning fossil fuels is extremely destructive.

Replacement of fossil fuels with other energy sources would greatly decrease the amount of CO_2 released into the atmosphere, but this will take many years and require a quick ripening of alternative energy technologies. As a result, even though many nations have agreed in principal to stabilize carbon emissions, no one knows how to do it, so the levels of CO_2 in the air continue to increase.

In view of the global risks, making rules no one has any intention of following and tinkering with unproved technologies while there is a continuous rise in greenhouse gases, is irresponsible and unacceptable. This is a problem that must be approached immediately and the best way to approach it is to find a way to store more carbon.

Many experts have pointed out that both afforestation and reforestation help remove CO_2 from the atmosphere. While a mature forest produces, through rotting biomass, almost as much CO_2 as it absorbs through photosynthesis, a young forest steadily converts more and more CO_2 to biomass. Tracts of forest store as much as two hundred tons of carbon per hectare, which is the equivalent of taking seven hundred and fifty tons of CO_2 out of the atmosphere. A fast growing forest can absorb as much as ninety tons a year per hectare.

This does not mean that rare old growth forests should be felled and new tree plantations put in their place. Old growth forests have incalculable value as reservoirs for biodiversity and felling them would destroy this, as well as, release immense amounts of carbon as the biomass rotted or was burned. There is ample room for new plantings all over the world.

The potential of afforestation is often underestimated because only the CO_2 absorption of the newly planted forest is considered, not the long term carbon storage capacity of the forest. To bind and store all the excess carbon being released through burning fossil fuels would indeed require millions of square kilometers of new forests. In a world where much of the forest cover has been destroyed, there is much depleted, eroded and damaged land on all continents. Restoring these areas might be the most effective and beneficial way to approach the problem.

Large scale afforestation efforts could replace lost vegetative cover and restore the topsoil, renew the water economy of degraded areas and increase the area's productivity. Projects such as these can be organized on a provincial, state or at the national level and be designed to meet the needs of the local people and the local ecologies. Another very important benefit from such endeavors will be work for the unemployed and underemployed, a population percentage usually of single digits in wealthy countries, but in poorer areas the unemployed make up one third to one half of the adult population.

Approximately five million square miles of replanted forest could store all the new carbon emissions for the next 30 years...a mammoth and seemingly impossible undertaking until one considers how much of the earth's surface has been destroyed by overgrazing, ruined by erosion, or worked out by unsustainable annual cropping. It is these areas which should be replanted in trees. The benefits besides greatly increased carbon storage would be numerous.

The lack of firewood in the third world is already a problem. Greatly increasing the tree cover in desolate areas would mean that the downed wood could be used for fuel or that trees could be coppiced. While wood burning also puts CO_2 into the air, this type of pollution in no way compares with the scope and intensity of the burning of fossil fuels in the transportation and industry sectors.

Reforestation and afforestation could supply sustainable building materials. Planting trees that can be coppiced, that is cut back and allowed to regenerate from the root is an excellent way to produce large amounts of usable wood. This ancient system was abandoned for the truly destructive method of clear cutting, or harvesting an entire forest with the resulting waste of secondary species, destruction of the undergrowth, destruction of habitat and loss of biodiversity. The reasons were purely economic, in the most derogatory sense of the word, for it is cheaper to come in and harvest a forest all at once, even though it often ruins the land, than to pay the kind of skilled and knowledgeable employees needed to manage a forest property and get the most of it without destroying the resource base. This should be quickly rectified as massive reforestation efforts could save the world we know by providing at least a temporary solution to the overload of the carbon cycle.

Trees are crop engines without compare, with literally acres of photosynthesizing surface area above ground, a net of soil building and soil holding roots below ground, and a harvest index no annual plant can equal, but because trees are less suitable for factory farming than plow and field crops, their utilization and development has been neglected throughout the 20th century.

Trees should be planted to supply wood as was already mentioned, but the useful products from tree crops are innumerable. These great perennial plants which do not need the plow or chemical fertilization, protect the earth and underpin the soil's water storage capabilities, and can supply much more than firewood, charcoal and building materials.

Trees can provide animal feed, carbohydrates for human consumption, edible oil and material for making ethanol and alcohol. Trees can supply fruits, nuts, fiber, sugar, medicines, teas, spices and raw materials for Integrated Pest Management (IPM). Trees can be hosts to valuable fungi, pastures for micro-livestock, sources for rare substances, and engines for pulling up minerals from the subsoil to enrich the topsoil. Trees can mitigate the drying or freezing winds, return the rains to areas we have turned into desert, break up our compacted and sterile soils which have been destroyed by continuous annual cropping, and restore the soils to fertility.

Importantly, some trees also fix nitrogen. The nitrogen cycle is just as critical as the carbon cycle to global health but the problem of misused nitrogen, which seems to have less direct influence on climate, is generally ignored.

There are important points concerning the nitrogen cycle which should be more widely known. Nitrogen is the most common gas in the atmosphere. It is required for plant growth. Nitrogen is the main building block for the construction of proteins, enzymes and other structures necessary for life. Free nitrogen (N_2) is all around but only a few organisms can use it as it is. These organisms can "fix" nitrogen so that other organisms can also use it.

The nitrogen "fixers" are vital to the first part of the nitrogen cycle in which free nitrogen from the atmosphere is turned into something that other creatures can utilize. Most of the nitrogen fixing organisms are bacteria that live in the roots of some plants or inside the plants themselves. They convert nitrogen into forms such as nitrate (NO_3), nitrite (NO_2), ammonia (NH3), and ammonium (NH4). These forms of nitrogen can be used by plant cells. The plants are then eaten by animals.

The second part of the nitrogen cycle deals with the breakdown of plants, animal waste and animal matter and the release of the nitrogen back to the air.

This is also accomplished by other bacteria, this time by specialized organisms that decompose the nitrogen based compounds which are so vital for life and turn them into free nitrogen again.

Unfortunately, besides being overlooked, the nitrogen cycle has also been overloaded. The production and widespread use of nitrogen based fertilizer and NH4 emissions from animal manure are two sources of the surplus nitrogen in the air and water. Even more surplus nitrogen enters the soil as rain washes down the pollution from burning fossil fuels, too much to be used by plants or de-nitrified by decomposing bacteria. The surplus nitrogen often does harm, for example the nitrogen heavy run-off waters that flow into rivers and then to the sea encourage huge algal blooms and die-offs. These events lead to hypoxic dead zones by the deltas of major rivers and raise water temperatures wherever they occur. The extremely warm waters in vulnerable areas often exacerbate storms.

Taking nitrogen out of the atmosphere should also be a priority, luckily this can also be done by planting trees, a special class of leguminous trees which grow quickly, improve the soil by fixing nitrogen and which protect and encourage annual field crop production even as they do so. Also able to help restore the nitrogen balance, are many pioneer plants that grow in association with the nitrogen fixing bacteria Frankia. These include the myrtle, the buckthorn, the alder and the birch.

Instead of converting forests to farmland, we should be doing exactly the opposite. We should be converting areas planted with damaging, uncompetitive, nutrient- greedy, labor-and-input intensive plow crops into great forests of food-bearing, resource-producing trees. Too many people, as they think of forests, see in their minds eye a forest of pines with maybe a bear and picnic table in it and nothing to eat in sight.

Forests which are planted in the 21st century should be forests of stock feed trees so that most of the wheat, corn, rice, and soybeans we grow can be used for human consumption and not end up as animal feed in a Concentrated Animal Feeding Operation (CAFO).

Forests planted in the 21st century should be food forests, planted to nourish us with fruit, nuts and edible leaves. Forest should also be planted to produce medicines, spices, rare substances and teas. Forests should supply firewood fuel energy and building material. Finally, forests should be planted to produce carbohydrates so that we may stop spoiling the world to grow our cereals and grains and confine their cultivation to areas suitable for this type of agriculture.

While we reap these benefits and collect these harvests, the trees we plant will store the excess carbon and fix the surplus nitrogen, cool and protect the suffering earth, redeem the barren, dying, and eroded landscapes found all over the planet.

Why then, do we not plant trees to save our world?

Is it too easy a solution?

Chapter 3
Fertilizer Trees

In the 2Oth century, fertility came out of a sack. The sack was purchased by the farmer and the contents spread on the land or put in the irrigation water. The result was bountiful crops and 20th century agricultural scientists rejoiced and did their best to spread this kind of farming all over the world.

However, by the end of the 20th century the picture did not look so rosy. The chemical salts in the sack, heavy with P, K and N, tended to wash out with rain or excess irrigation water, getting into the air, polluting rivers and making dead zones in the oceans where the rivers flowed out to sea. The chemical salts also tended to destroy microorganisms in the soil and make the soil unfriendly to beneficial insects and worms so when the soil became depleted, there was no biological base for its renewal. The contents of the sack also cost too much for many farmers to afford, used too much non- renewable energy while it was being produced, and produced foodstuff of great appearance but of poor quality and dubious nutritional value. In short, the use and overuse of chemical fertilizer created a bubble of abundant, but not-so-good food which is unsustainable in relation to the world's resources of water and soil.

The very materials used to enrich the soil have now proved to actually impoverish it. The use of chemical fertilizer, thought to be a solution, has proved instead, to be another problem. One we are going to have to solve as we try to untangle the vast untidy knot of unwise practices and the difficulties they have caused in order to have success in the search for sustainable methods of food production.

It is for this reason organic farming is on the rise again, after many years of being relegated to the fringe. But, to put conventionally farmed land in shape for organic farming takes years of intensive work to remove objectionable chemicals and much composting, inoculation, and mulching to restore what ought to be naturally in the soil. It is difficult and expensive to do and the cost of organic produce reflects this.

For much of the depleted, spoiled and worked-out land in the world, it is simply too expensive. The local farmers cannot afford it. Most of the governments will not subsidize the process, even though many governments subsidize chemical agriculture, simply because they mistakenly see a return to organic methods as a step backward.

If fertility is to be restored it must be restored in another way and the most logical way to do this is to examine the contents of that expensive sack of concentrated and toxic fertility and ask, "what was in it?" Most likely the sack contains the aforementioned minerals Phosphorus, Potassium and Nitrogen, or P, K, N, in various compounds and ratios.

Knowing this, how can we restore or create fertility in a less damaging way?

Nitrogen: the Most Commonly Missing Element

The number one ingredient of almost all widely used chemical fertilizers is Nitrogen (N). It is the nutrient which is most commonly deficient in poor soils, depleted soils, and marginal soils. So there are good reasons to try to increase the nitrogen content of crop land. The vegetative growth of leaves stems and roots is particularly dependent on nitrogen. Nitrogen deficient plants are often small stunted, discolored and generally unhealthy.

Molecular nitrogen (di-nitrogen, N_2) makes up four fifths of the atmosphere but in this state it is metabolically unavailable to higher plants and animals. However, nitrogen in this form is available to some microorganisms through a process called Biological Nitrogen Fixation in which the atmospheric nitrogen (N_2) is converted into ammonia (NH_4) by the enzyme nitrogenase in nitrogen fixing bacteria.

Microorganisms which have this rare nitrogen fixing ability are called "diazotrophs." Diazotrophs can be divided roughly into two groups, free living diazotrophs and photosynthetic diazotrophs. The free living diazotrophs basically fix nitrogen for themselves, usually using a chemical energy source. Photosynthetic diazotrophs use light energy to accomplish the same thing.

Associative nitrogen fixing microorganisms are those diazotrophs which live in close association with plant roots, either in the root zone (rhizosphere) or in the plants themselves. These diazotrophs obtain energy from the host plant to fix nitrogen and the host plants benefit from the nitrogen the microorganisms have made available.

While diazotrophs make a modest contribution of fixed nitrogen directly to agriculture and forestry, the amount of nitrogen needed for cropping is so great, the contributions of free living diazotrophs are not adequate. So the association between organisms such as rhizobia and legumes are used to provide large quantities of nitrogen to enrich nitrogen poor soils. The symbiosis between legumes and nitrogen fixing rhizobia takes place within specialized nodules on the roots or sometimes on the stems of the plant. A similar symbiosis occurs between certain types of woody plants and the diazotropic actinomycete, Frankia. In both cases, the host plant supplies the energy to the diazotrophs and the diazotrophs reduce atmospheric nitrogen to ammonia. This ammonia is then transferred from the bacteria to the plant for the plant's nutritional needs.

Nitrogen, in the form of ammonium ions, is also added to the soil from the decay of plants and animals which have benefited from the fixed atmospheric nitrogen in their life cycles. The nitrogen which originates from decomposing animals, plant residues, and also dead micro-organisms, is in essence nitrogen which has already been fixed and is now cycling through the ecosystem. Usually, there is not enough of this nitrogen, even in rich soil, to support the kind of intensive cropping which has become the modern norm. It is this nitrogen deficiency which has led to the widespread use of synthetic nitrogen, on one hand, and also, an urgent desire to find a way to replace the synthetic nitrogen with an enhanced biological process, on the other hand.

It is important to find a replacement for synthetic nitrogen because non-biological fixing of nitrogen is a process which requires large amounts of energy. It is primarily done by the Haber-Bosch process which consumes extensive amounts of fossil fuel. There is some nitrogen fixing by the effects of lightning, of course, but this is a natural process which fixes about ten million metric tons a year and has not changed over time. Artificial nitrogen fixing, however, has increased from three and a half tons in 1950 to ninety-one million tons in the year 2000 and to one hundred and twenty million tons in 2005. One and a third tons of fossil fuel is currently needed to fix one ton of nitrogen, an unhappy ratio which reflects the unsustainability of this kind of fertilizer use.

There are, however, many other significant reasons to seek alternatives to the manufacture of more synthetic nitrogen. These include the unhealthy, water-filled, growth of plants which are given too much synthetic nitrogen in cropping formats, the acidification of the soil by over-application of fertilizer in the form of chemical salts, and the extremely negative effects of fertilizer run-off throughout the riparian areas and down into the sea. When large amounts of nitrogen collect in a body of water, eutrophication can occur. Eutrophication occurs during an algal bloom and die-off such as the "red tide", a process which can suffocate all life in the water as the decaying algae uses up all the available oxygen dissolved in the water.

Acid rain also has its origin in the overloading of the nitrogen cycle. Nitric oxide in the air, which comes from burning fossil fuels, can react with water vapor to form nitric acid which is brought down to earth in the falling raindrops.

As mentioned in a previous chapter, we have no clear idea of what the effect to the environment will be from the disruption of the nitrogen cycle in the long run. This provides a significant incentive to find alternative ways to get fixed nitrogen into our agricultural systems. This is one reason why the associative diazotrophs, such as Rhizobia and Frankia, look more and more interesting in the context of the search for an ecologically friendly way to fix nitrogen. These bacteria fix nitrogen for their host plants in sufficient quantity to eliminate the need for chemical fertilizer. As green manures, nitrogen fixing crops are often turned into the earth, supplying the fixed nitrogen and the nutrients released by the decomposition of the plants. It is estimated more than three hundred and forty kg. of nitrogen per hectare can be fixed by fields of alfalfa or other legumes such as vetch and clover.

Enriching the soil with compost and animal manures, which break down and release ammonium ions, is immediately effective and while it is more labor intensive than spreading synthetic nitrogen, it is much less likely to disturb the pH of the soil.

It has been found the nitrogen content of soil improves notably when the soil is left fallow. Also, there is less loss of available nitrogen when crops are rotated. Together these practices do supply enough nitrogen for the healthy growth of agricultural crops and they must be encouraged simply because they have very few negative effects as compared to the deleterious effects of the application and the production of synthetic nitrogen.

Nitrogen Fixing Trees

Nitrogen fixing trees (NFTs), the hosts of nitrogen fixing bacteria, can do much to restore depleted and ruined agricultural lands. The high nitrogen content of their leaf litter makes them particularly valuable. The trees themselves have the advantages conferred by their own fertilizer production, making a "home" for symbionts and thriving in extremely challenging areas. When used as living fences and windbreaks, they protect and nourish grain and vegetable crops in improved fallows. Burned to fertilize the soil with their ash in the traditional "gum garden" cycles of the arid zones of Africa, they confer enough fertility on the barren earth to nurture food crops for several seasons.

Most NFTs are exceedingly hardy plants with well developed root systems which break up compacted soil, and hold and stabilize dunes. These are the plants which can do much of the work of restoring damaged and worn out soils simply by being planted in areas which have been abandoned after conventional agriculture is no longer profitable there.

The Sesbania, Gliridia and Pterocarpus families of trees have been used successfully for soil regeneration and fodder production in sub-Saharan Africa. The leguminous Acacia family has over a hundred species which are suitable for soil enrichment, browse and fodder production, and also includes species which produce seed pods which can be ground into high-protein flour extenders. The Prosopis family of trees contains a wealth of useful NFTs which are deep rooted, swiftly developing, multi-purpose species. These are pioneer plants, which hold tenaciously to life in harsh, hot, and dry conditions, become islands of food, shade, and safety for other plant and animal life.

In mixed tree fallows in Zambia where NFTs have been integrated with maize, the soils have changed physically, with better water infiltration and holding capacity, reduced crusting, and greatly increased organic matter. A study, Occasional Paper 05, The Impact of Natural Management Technologies: Fertilizer Tree Fallows in Zambia, published by the World Agroforestry Centre [aka International Centre for Research in Agroforestry (ICRAF)], documents the work of local farmers and a research team from the ICRAF in introducing these trees and carefully tracks the changes to soil and cropping systems.

The many wonderful benefits of tree fallows are listed in the study, starting with increased yields of maize; production of poles, stakes, and firewood; source of fodder for livestock; improved opportunities to grow vegetables; suppression of noxious weeds and insects; provision of shade; additional income from sale of seeds and saplings; reduced run-off; windbreaks; and enhanced biodiversity. The project was so successful; the number of farmers planting NFT in Zambia's eastern province jumped rapidly to 77,000 by 2005 and is still climbing. On average, the value of the maize crop increases by 20% and the other benefits brought by the NFT tree fallows added another 10 to 15% to the participant's family income.

In tropical areas, Leucaena trees have been used extensively for soil regeneration, fodder and wood production. Called "the Multiplier Tree" and originating from Central America, it is useful as a nurse crop and green manure as well.

In Australia mixtures of eucalypts and acacias are being planted to protect and improve depleted topsoil.

There are NFT species which grow in cold and temperate zones as well including some of the most common trees. See Chapter 7 for lists of suggested species for various climatic zones.

The Nitrogen-Fixing Bacteria

Rhizobia

In one of the biological world's most interesting examples of mutualism, the host plants of some nitrogen fixing bacteria "invite" the bacteria to congregate in the rhizosphere around the plant's roots by releasing a variety of chemicals into the soil through the roots cells. Some of the chemicals released encourage the growth of the bacterial population in the rhizosphere.

Reactions between the bacterial cell wall and the root surface take place, which allows the bacteria to recognize the proper host plants. The bacteria then attach themselves to the root hairs of the "inviting" host.

The rhizobia bacteria, once bound to the root hairs, make the next move. They excrete compounds called "nod factors." These compounds stimulate the root hairs to curl. Rhizobia then invade the root through the hair tip where the nod factors induce the formation of an infection thread. Oddly enough, the infection thread, a sort of super-highway for bacteria, is constructed by the plant, not the bacteria, just as the plant constructs the maze of enclosed tubes which make up the root nodule. In these nodules millions of bacteria live. It is in the symbiosomes, specialized cells of plant membranes which interface between the bacteroids and the living tissue of the plant, where the fixation of nitrogen takes place.

Frankia

Bacteria of the genus Frankia belong to the class actinobacteria. These bacteria were originally linked to fungi because of the mycelium like filaments many of them form. Twelve species of Frankia are recognized, nitrogen fixing bacteria which live symbiotically with a large variety of dicot plants. These plants, with their bacterial symbionts, are responsible for fixing about 15% of the biologically fixed nitrogen in the world. Alder and myrtle are two of the pioneer plants which live in association with Frankia. The hardy Casuarina, a heat and salt tolerant tree from Australia, also lives in association with Frankia. The leaf litter of these three trees is extremely rich in nitrogen and the trees stay green for long periods of time, even when surrounding species are remobilizing their nitrogen and dropping their leaves. Some of the other trees and plants which benefit from Frankia nodulation are birches, oleasters, buckthorns, eleagnus (autumn olive), Shepherdia (buffalo berry), and Hippophae (sea buckthorns).

The Frankia bacteria also cause root hair deformation. Then the bacteria penetrate the cortical cells and induce the formation of nodules. The nodules are colonized by vegetative hyphae (mycelium) which differentiate into round, thick walled structures called "diazo-vesicles" where reductive nitrogen fixation takes place.

Other Symbionts that Improve Plant Nutrition

There are other organisms that fix nitrogen. Cyanobacteria are examples of such organisms. These blue green algae are found in many terrestrial and aquatic habitats and their thick mats of biomass are sometimes used as food for animals or fertilizer for the fields.

The aquatic fern Azolla is the only fern which can fix nitrogen. It does so by a symbiotic relationship with a specific species of Cyanobacteria (Anabaena azolle). Azolla ferns are often planted in fishponds and dike/pond integrations to supply nitrogen for the cultivation of rice, lotus and other water plants.

Leafy liverwort (Porella navicularis) is an abundant epiphyte in the Pacific Northwest. These plants benefit from their association with Cyanobacteria called Nostoc which lives on the plant's leaves. Nostoc also forms associations with other bryophytes. Nostoc also can be found in lichen where they occupy specialized pockets within the plant. These in-dwelling bacteria are called cephalodia.

Cycads are ancient plants which are considered the forerunners of modern palm trees. Cycads are the only gymnosperms or naked seed plants which form root nodules and are also the only vascular plants which team up with cyanobacteria to fix nitrogen.

Mycorrhiza is fungus which lives in a mutually beneficial association with plant roots.

Micro-symbionts are symbiotic fungi and bacteria which improve plants' ability to absorb nutrients. It is estimated, in a tablespoon of ordinary soil there are thousands of miles of fungal filaments, unseen, but a crucial living bridge between the mineral world and the plant world. It is these invisible links which are most put at risk by chemical agriculture.

These examples of mutualism are encouraging and they point the way towards a better method of introducing nitrogen to the soil than using synthetic nitrogen while keeping the elements of the soil in balance.

Potassium

Potassium is the second element in the sack of fertilizer, usually given in the relatively benign form of potassium nitrate (KNO_3) or in the nastier form of potassium chloride (KCl). These forms of potassium, besides being expensive, tend to leach out quickly in coarse soils and end up polluting watersheds and forming hypoxic zones. These are two good reasons to find a more ecologically friendly way to put potassium into the soil and to keep it there.

The major role of potassium in living organisms is osmotic control. It is important as an enzyme activator in plants and it is needed to facilitate membrane permeability and the translocation of sugars. Potassium is also required for photosynthesis, fruit formation, disease resistance, and protein formation. Potassium builds strong plant structures but it is not a permanent fixture of plant tissues. Instead, it passes in and out as needed by the crop. It is often translocated to root and stems during ripening of the primary products. Thus, potassium is readily available from crop residues. Very little potassium is needed if crop residues are returned to the soil. However, continuous cropping or the harvesting of hay or silage without returning potassium to the soil can quickly induce K deficiency.

Potassium is taken up, retained and excreted by living organisms in the form of a positive ion. In animal manures and decaying plant matter, the potassium is not fixed to other compounds and so is readily available to growing plants. This is sometimes called "soil solution K." The K in solution can change form and rest in negatively charged soil sites. Potassium moves back and forth between the state of exchangeable potassium and potassium in solution.

Potassium also becomes available by the weathering of clay and rocks but this is a slow process. Powdered basalt, green sand and clay minerals can be applied to the soil to correct potassium deficiency, but, the addition of organic material and the use of green manures will aid in potassium storage and keep potassium from leaching away. Healthy soil with lots of organic matter and abundant sites which are negatively charged can both fix and release potassium, thereby allowing it to interact dynamically with calcium and magnesium and be taken up by the plant.

Legumes and green manure crops are especially good for the balance of potassium in the soil. Leaves from deciduous trees are rich in potassium. Dry leaves from nitrogen fixing trees and from most windbreaks are good sources of potassium. Potassium in leaves and crop residue is generally 1-4 percent of the dry weight, enough potassium for a stable supply to the current crop if retained in the form of compost or humus.

Phosphorus

Phosphorus is the third element in the sack of fertilizer. Ironically, most forms of phosphorus, including those in commercial fertilizers, can not be taken up directly by the crop plants, even though phosphorus is very important in plant cell division and plant growth. This makes P a difficult nutrient to manage.

Although there is a good deal of phosphorus in most soils, it is often in an unavailable form. Phosphorus is bound in acidic soils to iron and aluminum. In alkaline soils it combines with calcium. Even with a neutral pH, phosphorus readily becomes immobilized. Since it becomes inactive so easily, it is difficult to keep enough usable phosphorus in the soil. Phosphorus anions can be trapped in clay humus complexes and be lost through erosion. It is another of the nutrients which can end up in eutrophic and hypoxic zones. Stopping soil erosion is one of the best ways to conserve phosphorus.

Another method to conserve phosphorus is to restore back to the soil the micro-inhabitants eliminated by chemical applications of fertilizers. It is possible to add powdered rock phosphate or rock dust to avoid phosphorus deficiency but the release of P to plants depends on biological activity in the soil. Soil acids produced by bacteria and mycorrhizal fungi act upon soil phosphorus and makes the phosphorus become available to plants. Therefore, phosphorus availability is not just a matter of supply and demand, but also, of maintaining high levels of biological activity in the soil.

Some plants produce acidity around their own roots which assists in the uptake of P. This is true of most trees and plants which fix nitrogen, some palms, rapeseed, buckwheat and many pioneer bushes and shrubs.

Replacing the Fertilizer Sack with Something Better

In the case of each of these major nutrients there is a sustainable less expensive alternative to the continued application of chemicals. In some cases, the same methods and organisms which promote increased nitrogen in the soil will also help solve the problems of potassium and phosphorus deficiencies. The methods include mulching composting, crop rotation, and the application of green manure.

The helpful organisms include trees, shrubs, some cereals, and leguminous plants which encourage the activity of countless tiny soil enriching and producing organisms. These organisms and practices are less costly in both the long and short term than the fertilizer in the "sack". They improve the agricultural and ecological system on many levels, as they create fertility by transforming P, K, and N into forms other living organisms can use. They are instrumental, as well, in retaining these forms in the soil.

The chemical sack of fertility should be relegated to history. It was an unsuccessful experiment and one which has proved to be too much of burden for the farmer, the agricultural system and the earth.

Chapter 4

Growing Bread on Trees

Bread, like money, does not grow on trees. Correct? But what if it did? What if carbohydrates equal, or even superior, in quality to those produced by cereal grains came from trees? Would the kind of agriculture we do change for the better?

I think it would.

We would plant more trees, cultivate more trees, and receive larger yields per acre while doing little ecological damage. We might also use trees to regenerate areas which have been worked out from previous improper agriculture since this would be profitable, as well as wise.

Four Approaches and a Fifth Possibility

There are four main approaches in which bread can be "grown on trees" and three of them would move agriculture much closer to sustainability.

Extend Flour with Tree Products

The first would be to cultivate trees whose products can be used as flour extenders. In this case, much nutritious substance and protein could be added to local bread. This can be done relatively quickly, as many tree products which can be used in this manner have already been identified. In some places, those trees which produce these products can be introduced and planted. In other places, the appropriate trees already exist and it is simply necessary to gather the products and use them.

Bread Trees

The second is to cultivate trees which produce seeds or products which can be made into bread. This is much more difficult than the matter of flour extension. Flour must be created, not merely stretched or enriched from the products of these trees. However, modern bread is a "raised" product containing gluten, this kind of bread from tree products is difficult to achieve as there are currently no "wheat trees" known. Still, raised breads can be produced from materials which have little or no gluten in them, such as corn meal, by forcing air into the dough, malting, fermentation or using a leavening agent which does not require gluten to act.

Stop Feeding Bread to Animals

The third strategy would be to replace much of the grain fed to animals with tree products. This would serve several purposes, the first being grain products would be used to feed people instead of dairy or meat animals. A slackening of demand for cheap cereals to fatten animals would mean less strain on the agricultural systems of the world. Other benefits from this strategy would be an increase in free-range animal husbandry with fitter and healthier animals and better milk and meat. Less medicines and antibiotics would be used since the animals will no longer be kept in crowded feedlots. Also, this would be a more ecologically "friendly" use of land since worked out and marginal areas could be regenerated by planting nitrogen-fixing, stock-feeding trees.

Engineer a "Wheat Tree"

The fourth strategy would lie in the realm of genetic engineering (GM). Genetic engineering has been a disappointment when viewed as a method for increasing the food supply. Trendy and purely commercial genetic modifications have been made in crop plants. Many of these modifications have proven to be unsustainable and dangerous. Other modifications are worthless in the long run, such as "built-in pesticides" which only develop resistance all the faster in insects than the sprayed-on pesticides ever did.

Yield drag in modified crops, lawsuits from spreading GM material, soil contamination by GM organisms, and recalls of foods made from crops unsuitable for human consumption have all made farmers and consumers justifiably suspicious of every GM strategy. However, genetic modification could be put in the service of a wiser kind of agriculture, if a few simple and logical rules were followed:

1. Do not modify food plants with genetic material from organisms which are inedible.
2. Do not modify food plants with the genes from humans, animals, insects, or anything but other food plants.
3. Make all GM plants sterile so GM genetic material does not spread to related plants by cross- pollination.
4. Test every new GM plant as carefully as if it is a new and unknown plant.
5. Do "pharming," which is the production of medicinal material by crop plants, in closed formats, with all products labeled and registered as carefully as prescription drugs.
6. Pass legislation to insure the person or company who is making the money from GM organisms is the one responsible for cleaning up the mistakes, not the general public.

If these rules could be enacted and enforced, GM strategies would be much less dangerous to the global commons. The risks of GM contamination of crops and crop relatives would be greatly reduced and the risk of inedible chimeras such as Starlink corn and the GNA potato getting into the food chain would be zero.

Under this system of rules, it might be possible to make a few modifications which would really change agriculture for the better. Why not modify a few copiously producing nut trees with a genetic sequence which produces gluten? These modified trees, also made sterile during the modification process, could then be safely tested as producers of raw material for making bread. If successful, then bread could really be grown on trees, rather than from the weak and fragile, over pampered, annual grasses which currently produce our grain.

The Potential of Perennial Grasses

A fifth strategy for a more ecologically friendly way to produce bread crops, the cultivation of perennial stands of grass, is already being tried in Africa and on the American prairie.

These five approaches are either in development or being seriously considered because the demand for grain crops is so high and we have changed so much of the earth's surface to produce them. We are literally mining the earth to produce our daily bread.

Now that we have discussed what might be done to "grow bread on trees," let's look at how these five approaches might be accomplished.

Nuts and Loaves

1. Stretching a Loaf: Flour Extenders and Flour Substitutes

Actually, many of the flour substitutes and extenders from trees have become known to humankind because of periodic grain shortages throughout our history. While many kinds of edible and inedible materials were used during famines to "stretch a loaf" (including sawdust), this section is confined to those materials which improve the extended flour nutritionally or replace a fair percentage of it in the baking process.

The Multipurpose Mesquite

One of the best tree products for this purpose is the ground seeds and pods of the mesquite tree. Mesquite is already known for its value in animal feeding. The tree is also cut or trimmed for high quality smokewood for barbecues. It is considered a very valuable tree for grazing and for honey production. Mesquite gum is a fair substitute for gum Arabic. However, this hardy tree from the Prosopis family also produces copious amounts of pods which have high percentages of protein and sugar. These pods and seeds can be milled into flour.

The mesquite pod is already used as food in the Central Mexican Plateau region, also known as the Mexican Altiplano. There the mesquite pod is consumed fresh, boiled in its own syrup, or ground into sweet flour called "pinole." They can also be made into candy (queso or piloncHillo) or boiled into a thin porridge with water, milk and corn meal. The mesquite pod can also be used to prepare a beer-like alcoholic drink.

Mesquite pods are ground into meal in a variety of ways from the home made mortars and pestles to advanced processing equipment which separates the pods into cotelydon, seedcoat, endocarp and exocarp and makes a different product from each part. The most common way of making mesquite flour for human consumption is to sort and air dry the entire pod before crushing it in a hammer mill.

Mesquite pods are high in fiber, 17% to 30%, and low in fat, 1% to 4%. The protein in the seeds is similar to soybean protein and the protein in the outer pod is similar to the protein in whole rice or barley. Sugar content of mesquite pods can be as high as 41%, or as low as 13%, but the sugar is mainly fructose which the body can process without insulin. Mesquite meal contains a high percentage of soluble fiber. The ground pods are high in lysine and a good source of potassium, manganese, and zinc. There is a good deal of variability between individual seedling trees and even more variability between mesquite species.

Used to extend flour, mesquite meal is blended in combination with other flours. A typical recipe uses 30% mesquite meal to 70% rice or wheat flour. It is especially good in cakes, cookies, bread, porridge and meal replacement drinks, as its slow absorption, and the resulting flattened blood sugar curve, staves off hunger and supplies a steady energy level to the body.

Carob

Carob flour, a fine ground meal produced from the pod of the Ceretonia siliqua is valued because it does not contain gluten. This makes carob flour suitable for producing baked goods for sufferers of celiac disease, wheat allergies and some people who suffer from diabetes. Carob flour has a moderate amount of protein, no cholesterol, is low in fats, and high in fiber. The major mineral supplied by carob flour is calcium and the major vitamin is Niacin. However, 89% of the food energy is supplied by carbohydrates and sugars, so it may not be suitable for some diets.

Carob flour can be added to other non-gluten flours at a ration of 1 part carob to 3 parts non-gluten flour. It can be used to extend wheat flours at the same ratio. Carob is often used as a chocolate substitute or a substitute for coffee for those who must avoid these substances. In North Africa, it is an ingredient of choice in baked goods because it behaves as other flours do in cake batters and imparts a special texture and heartiness to the final product. In Cyprus, the carob pods are processed into syrup. Carob flour burns easily when it is heated because it contains almost no fats. Its starch granules absorb moisture and rupture when heated, releasing a stabilizing gum. Carob flour is rarely used on its own but usually mixed with other flours, sugars and fats.

The ripe pods, shiny brown, and nick-named Saint John's bread or "boxers", can be eaten straight from the tree, but the small hard seeds, uncannily regular in size, are usually discarded. A steady diet of carob pods is not suitable for man or beast, as too much carob intake is bad for human digestion and limiting for animal fertility. The recommended ration for carob in the diet is currently, in both cases, not more than 25% of total food intake.

2. The Promise of Bread Trees

These are trees with seeds which are suitable for making bread and porridge. Very little has to be done to the seeds, except leaching and milling. An especially nutritious flour and meal is produced from them which is valuable when used to extend flour, but of such a high quality, bread might be made entirely from these seeds.

The Oak, the bread tree of antiquity

Literature and fiction about prehistoric culture often remake our ancestors in our own image. However, we are never going to know exactly what they thought and felt, how they worshipped, nor how they dealt with all the problems and troubles to which today's humans have inherited. We even know less about what they ate and how they hunted than we thought we do. Hundreds of sites have been dug up and the artifacts which were discovered have been investigated. Theories have been constructed and demolished by the archeologists who have examined these prehistoric sites and prehistoric middens (A "midden" is a mound or deposit containing shells, animal bones, and other refuse which indicates the site of a human settlement. They are also called "kitchen middens").

The most common mental picture of a prehistoric man is that of a muscular fellow, dressed in skins, hunkered in a cave, roasting a hunk of meat from some extinct behemoth over an open fire. The understanding is emerging only recently that prehistoric humans probably ate a much more varied diet than was formerly thought. They seem to have consumed anything and everything from fish, snails, worms, insects, small animals, birds, large animals and even large carnivores. They also ate a good deal of vegetable matter, including leaves, shoots, roots, grasses, and nuts.

One particular class of nuts appears to have been the mainstay of many prehistoric cultures. These are the acorns, a nut rarely eaten in modern times, but one which may be the closest thing to producing "bread grown on a tree" nature has to offer. In William B. Logan's fascinating book *Oak, the Frame of Civilization*, he devotes an entire chapter to his premise of the acorn as a staple food, perhaps the most important food in antiquity. This explains why so many people, in so many different places, have either records or legends of acorns being eaten. It also explains why many people used them well into the 20th century as a special seasonal food or a delicacy.

Logan cites Greek poets and historians. The Greek poet Hesiod wrote in the eighth century BC, "Honest people do not suffer famine, as the gods have given them abundant subsistence: acorn bearing oaks, sheep and honey." Ovid in a work called *Fasti* explained that people once lived more or less on the acorns. "The sturdy oak afforded splendid affluence." Pliny described all the oaks in his part of the world and described their many uses. He declared, "Acorns are the wealth of many races even when they are enjoying peace." Pliny adds to this statement, "When there is a shortage of corn they (acorns) are dried and ground into flour which is kneaded to make bread. In the Spanish provinces a place is found for acorns in the second course of the table." The "second course" was a cooked or baked course, so the acorns might have been served as bread, porridge, or mush.

This also solves a mystery which has puzzled archeologists for over a hundred years: in areas with no history of grain cultivation, millstones and grindstones are found in archeological sites. If there was no grain to grind—then what being ground into flour? Apparently, in these areas, acorns were the staple carbohydrate food and whether the nuts were pounded up in mortars and pestles, milled between stones, or crushed in pits by smooth rocks, as was done by the Native Americans in California, the resulting meal was baked on hot stones as bread or cooked with hot water and flavoring as porridge.

The practice of using acorns was not confined to the Mediterranean region. The legendarily fierce Hebrew tribe of Dan reportedly made acorns into the "waybread" they took with them to war. Acorns are still dried and salted or sweetened to make snack foods in parts of Spain. Kurds and Turks still make foods out of acorn flour, spiced and scented or mixed with buttermilk. Acorn flour and acorn tofu can be found in Korean supermarkets. The California Indians based their entire system of food storage and cookery on the acorns of California's many species of oak. They turned the nuts into mush, bread, soup, crackers, and a fermented paste like poi. Oaks still grow easily and abundantly in the Mediterranean region, in France and England, in California and Oregon, in the cooler parts of Southeast Asia, in the mountains of China, in Turkey, and the Soviet Union.

Their abundance goes mostly ignored and unused. New York's Central Park, home to the world's fattest squirrels, has a serious disposal problem with oak leaves and acorns, discarding thousands of tons of oak "mast" yearly. This mixture of leaves and fallen nuts in other places and other times, would have been used to fatten sheep, goats and pigs. Now it is considered garbage, an expensive nuisance to be hauled away and dumped into a landfill.

Could the oak become an important food tree again?

For a brief time in recorded history the oak became a "bread tree" again. The Little Ice Age, a period wherein the population in Europe plummeted as wheat and rye based agriculture suffered through years of cool wet summers and winters when the ground froze solid, brought the acorn back into the pantry. Bread was made with the 50% addition of acorn flour. Acorn mush, flavored with herbs, became a staple food.

The famine in Sweden in the 1840's also saw a return to the use of acorn bread and mush. A large peasant population, farms which had been subdivided by inheritance laws into holdings too small to support a family and the same wet weather and short summers which brought on the "potato famine" in Ireland also brought northern agricultural systems into crisis.

A wave of emigrants emerged from Sweden and Norway, heading mostly for North America but settling as far away as Australia. The farmers who stayed home increased their holdings by acquiring land sold by the emigrants and eking out their bread grains and hay with acorns, oak mast, and the wild products of forests and meadows including hazelnuts, lichen, the roots of bracken, and ferns. Some parishes, lost half of their population to migration or starvation between the years of 1840 to 1875.

If acorns are good enough to eat when little food is available—then, why are they not used all the time? Part of the problem is the bitterness of so many acorn types. This bitterness has to be leached away by soaking the acorns in water. The acorns can be leached quickly by boiling them and changing the water several times. Or they can be leached slowly by burying them in the wet gravel by a stream so the slow action of cold water will take away the bitterness. Modern plant breeding could select a few sweet acorn varieties of oak and propagate them, but oddly enough, few people remember the acorn is excellent food until all their other food disappears.

The second problem with the acorn as a food is the leached acorns and acorn flour is almost tasteless. The flour smells good. Mush bread or tofu made from acorns is wonderfully nutritious and very filling. However, acorns must be cooked or mixed with something else to give them flavor. In a mush, soup, or stew, this is not a problem.

As a whole nut or made into bread, the flavorlessness of the acorn is unappealing. It is for this reason acorn nuts are toasted, salted, or sweetened, and acorn bread is flavored with herbs or seeds. This is also a problem which could be overcome by plant breeding and selection—or simply by seasoning the flour or adding it to another type of flour with a more conventional taste.

Among the trees of the many varieties of oak, there are individual plants which produce acorns with stronger, nuttier tastes. If this is what the potential consumers want, as regards to the flavor of acorn flour, trees with this quality would not be hard to replicate. What is missing is the will to use and cultivate the oak for food—unless the bread from the field fails. Then suddenly, it is remembered, "the bread of antiquity came from the forest."

Acacia, the bread tree of tomorrow

There is another tree even more ubiquitous than the oak, hardy, drought tolerant, and nitrogen fixing which lives happily in the most desolate areas of the world. This would be the acacia tree which graces with its rough trunks, thorny branches, and tiny green leaves, the sands and rocks of the most forbidding deserts.

Acacia seeds have long been a traditional food in Australia. Aboriginal women winnowed, parched, and ground the seeds of the Acacia aneura, turning the flour into both bread and porridge. If acacia seeds are going to be used for food for humans, the traditional knowledge and skills of these women will be very important.

Australia's Commonwealth Scientific and Industrial Research Organisation (CSIRO) held a workshop in 1995 which involved the Aboriginal community in seed collection, taxonomic studies, and the recording of traditional knowledge. Species trails in outback communities were planned and nurseries were established. Then seeds and trees were introduced to a similar climatic zone in sub-Saharan Africa. Some Australian acacias had already been introduced in the 1970's to serve as windbreaks, stabilize the soil, provide firewood, and halt desertification. The acacias introduced after the workshop were species specifically chosen for their potential value as human food.

Acacia seeds are splendidly nutritious. They are more dense and concentrated than most plant parts and contain significant percentages of carbohydrates, edible oil and crude protein. For example, the seeds of A. victoria contain 18% crude protein, a third more protein per gram than wheat varieties. The addition of acacia seeds gives traditional stews, pancakes, and starch dishes a welcome heartiness. While some acacias like Acacia colei, currently under experimental cultivation in Maradi, Niger, have seeds which can be added to sorghum and millet bread or porridge, other acacia seeds can be made into malted or unleavened bread. Some acacia seeds and pods can even be milled in the same manner as mesquite pods are milled.

So far A. colei, A. cowlena, and A. tumida seem the most appropriate acacia species for sub-Saharan Africa. These three species are easy to establish, fast growing, highly nutritious, and can be winnowed and gathered easily. They show no signs of toxicity or anti-nutritional characteristics. Other species may also be suitable and the search for edible acacias suitable for different areas and soil types in both Africa and Australia continues. Twelve species of acacias have been planted in provenance trails in Australia's most arid zone.

The introduction of edible acacias in Niger and Ethiopian, has yielded such promising results that five more African nations have started their own acacia projects.

The growing "Bush tucker" industry in Australia, which is built around native foods and traditional Aboriginal knowledge, has familiarized many people with the kinds of food eaten before the European settlers arrived. Acacia seeds were a staple of the aboriginal diet before the introduction of wheat flour. The Australian Native Produce Industries, one of the largest firms dealing in "bush tucker" has successfully used acacia seeds in the production of linguine, damper, pavlova, and ice cream.

It is assumed Aboriginals began to use acacia seeds as food as long as 20,000 years ago. Grinding implements of great age have been discovered all over the continent. There is no reason why the acacia tree, having been a bread tree for thousands of years in the past, should not resume its role and be a bread tree of the future.

3. The Price of Feeding Bread to Animals: Why This Practice Must Change

The major food crops of the world, providing more than 80% of the food energy for man and beast, are the cereals, the soybeans, and the potato crops. These are crops which must build themselves anew each season and in doing so require immense amounts of water, fertilizer and care. They are relatively weak plants on a competitive level and they can only be cultivated efficiently if they are planted on clean, flat, well-tilled earth, with all the clods broken up and possible competitors eliminated.

These are crops best suited for thick layers of topsoil and for alluvial plains. The new powerhouse of nutrition, the soybean, is cultivated in the same format as corn, cotton, rape seeds, and a dozen other mega-crops of the modern world. The substance of this kind of agriculture has not changed in thousands of years. It depends on plowing the land and making it bare, turning over the earth and making it vulnerable. Just this single practice is enough to cause the erosion of hundreds of thousands of tons of topsoil every year.

An uncomfortably high percentage of these ecologically taxing food crops are fed to animals. This includes corn, wheat, soybeans, and oil seeds. The animals do not need these foods. They are naturally equipped to digest grass leaves, browse, and stubble, and many other plant materials which keep them healthy—all substances people cannot eat.

Domestic animals, built to roam and graze in herds, are today confined to vast feedlots where they can hardly move. Highly dosed with antibiotics to keep them safe from the diseases of overcrowding, standing in their own droppings, and fed constantly, they fatten very quickly. After butchering, their fat-marbled anti-biotic contaminated, meat is fed to increasingly unhealthy and obese citizens of the developed world who suffer from clogged arteries and other diseases caused by eating the meat of unhealthy animals.

Meanwhile, there are many, many, people who cannot raise and cannot buy their daily bread. Unsustainable crops, misused land, unhealthy animals, and people who are starving to death or dying of surfeit—there is little about this system which makes any sense. The situation *must* be improved. The devastated, eroded, abandoned areas of the countryside must be replanted with species of trees which provide high protein pods and browse, thereby providing food for the animals and reforesting the abandoned land. Animals must be allowed to graze, browse, and roam, eliminating many of the diseases of overcrowding and the need for drugs and producing leaner meats and healthier products. We must drop the expensive practice of subsidizing the production of grain, so that, the farmers of the developing world can grow their own crops for human consumption and not have to compete with the flood of cheap corn industrially grown to underpin animal feeding.

This complex web of destructive practices has to be unraveled sometime. It should be done while there is still choice about how and when we change it. This system impacts very negatively on the global food web—but the collapse of this system would be unimaginably catastrophic. Willing modification towards sustainability will prevent much suffering and save many lives in the future.

4. Growing Bread on Trees

Should we turn wheat and corn into perennial plants? (Other staple crops have perennial types that are now being investigated)

Should we modify nut trees to produce flour for bread?

Are GM strategies necessary? Or can we find solutions in smart breeding?

Basic research, analysis, and evaluation, which has already been done, indicates it may be possible to achieve high yielding varieties of *perennial* corn and wheat. While the perennial corn plants are still in the experimental stages, perennial wheat, which has to be replanted only after the fifth year is already a reality. Perennial sorghum and rye seeds are also already available on the market and there has been much progress in breeding perennial versions of sunflower, flax, chickpeas and rice.

The current perennial wheat varieties are not yet equal in their yields to annual wheat, but they do show promise for challenging cropping areas because they eliminate the need for annual plowing and mitigate some of the other negative effects of cereal cultivation. The development of the roots is the key difference between the perennial and annual plant varieties, affecting the roles they play in soil conservation and nutrient cycling. Perennial wheat sends roots down three times as far as annual wheat, holding the soil tightly in place even on sloping land. This soil binding root net does not die off annually, making perennial plants very effective at controlling erosion. In some cases, perennial wheat has been demonstrated to be six times as effective as programs of improved management of annual crop wheat with crop rotation and mulching. Soil structure improves under perennial wheat plants and there is better water infiltration. Perennial wheat plants cycle nutrients more efficiently, resulting in less leaching of nitrogen into the watersheds. Perennial wheat crops use 90 percent of added fertilizers, leaving less to run off and pollute streams and ground water. Also, perennial wheat crops do not allow the easy growth of weeds and competing plants. As run-off trapping borders, swards, forage, grazing, and anti-erosion plantings, even a minor shift to the perennial varieties of these crops could have great benefits for agriculture.

Perennial stands of wild grasses are being researched. Some have yields in quality and quantity competing favorably with domesticated annual grains. Perennial grasses have long growing seasons. They are already in place and ready to resume growing when the spring rains come. They collect extra solar energy, a fact which is reflected in a good harvest index. They are an ecological blessing to challenging areas as they continuously cover the soil, stabilize that soil, facilitate rainfall infiltration, and enhance absorption of water which would otherwise run-off and cause erosion. The savings in energy, as these stands do not have to be plowed or fertilized, and the savings in edible material, as seeds do not have to be saved for replanting, make the possibility of perennial grain crops an exciting option.

One of the places these exciting possibilities are being studied is the Land Institute of Salina, Kansas. There Wes Jackson and his colleagues are attempting to revolutionize the way carbohydrates are produced. Their aim is not an improved wheat field but a self-planting, self-sustaining ecosystem like the prairie, but a prairie which produces edible foliage and edible grain. These innovative methods may lead the way to a much more sustainable way to produce our daily bread.

What about engineering a brand new crop—choosing a copiously producing nut tree like the oak, the beech, or the hazelnut and modifying it to produce gluten?

This is a bit of a long shot as it requires genetic engineering to be done with a level of skill which is not currently possible. It would involve moving a genetic sequence for the production of gluten which would function without any negative effects in the new organism and for this function to not be silenced by the epigenetic system of succeeding generations of this new organism—hopefully achieving flour that can be used in the making of raised breads.

This goal is not quite as strange as it sounds. Nut flours are very nutritious and have more protein than the flours of cereal grains. Sufficient oil is already present in ground nuts for baking. However, 100% nut breads are heavy, even denser than unleavened rye bread. The missing element is the elasticity normally provided by gluten which would allow the bread to be leavened and rise. This could be solved by the addition of gluten, but part of the appeal of bread made with substances other than wheat is the fact that most of these flours are gluten free. Rather than genetically modifying the tree products to make bread from them, it might be much cheaper and simpler to learn how to make raised bread from non-genetically modified nuts and seeds.

Much basic research must be done before GM strategies are considered as a solution. These are some of the subjects which must first be explored:

1. Current use of tree-produced carbohydrates
2. Evaluation of current carbohydrate crops by comparative analysis
3. Re- evaluation of "lost" carbohydrate producing tree crops and famine foods
4. Identification of possible "bread trees" among the palms
5. Bread trees and oil trees among Indigenous wild species and their possible domestication
6. Comparative harvest index exploration
7. New products and processes from promising bread plants
8. Profiles, qualities, and characteristics of bread making materials from alternative sources of carbohydrates
9. New food technologies that might be usable with perennial carbohydrate producing crops

Genetic modification to achieve the desired result is not an easy or a quick process. It has potential risks and negative impacts. Despite its reputation as a technological quick fix, GM strategies take as long as, or even longer, than conventional breeding. Also, while gluten is necessary for elastic dough, there are more ways than one to make raised breads.

For instance, nut flour can be malted, as sorghum flour is malted, to achieve a raised bread or dough. It can be puffed up with air, as is done when making rice-cakes. Corn flour has no gluten but cornbread was the respectable staple of the American South and Southwest for three hundred years.

If it comes down to a choice of adding new bread making possibilities by creating new plants by GM technology, or using new technology to make raised bread from veteran foodstuffs in new ways, then it is obvious that new food technology is both cheaper and safer and can put new products on the market while the GM advocates are still getting their gene guns out of their holsters.

New food technologies come into use rather quickly. One need only look at the development of the edible products derived from hemp seeds, for example. Hemp oil and hemp fiber are extremely useful material and they have been in use in many parts of the world for hundreds of years. The oilcake, however, has only been considered food for human beings for the last decade or so. Quite suddenly the oilcake has been identified as a valuable food and processed into ingredients for cookies, food supplements and power bars. If a fraction of the amount of time and money spent on hemp would be invested into making raised, or otherwise more palatable, breads from nut flours, those products would bloom just as quickly.

As we search for a way to grow bread on trees these are the questions that must be answered:

1. Can the perennial versions of annual crops move agriculture towards greater sustainability?
2. Can the products of wild and domesticated trees provide “bread” or a reasonable substitute?
3. Can perennial crops, both wild and previously domesticated, supply material to replace a significant amount of the grain used in the production of flour?
4. Can arboreal or other perennial crops which reclaim spent soils also supply browse and mast in sufficient quantity and quality to replace a significant percentage grain and cereal crops in animal feeding programs?
5. Can transgenic plants truly change current agricultural equations regarding the production of cereals and grain?

These questions can be answered by investigating new crops, lost crops, current crops, and wild species of perennial plants for the qualities and characteristics necessary for producing bread, bread substitutes, and animal feed. This research could improve the balance in the agricultural world and allow the cultivation of cereal and grain crops to stay in areas best suited to that type of cultivation.

The exploration of trans-genetic modification of arboreals for carbohydrate production, the possibility of turning maize into a perennial plant, and the serious examination of perennial grain producers also have bearing on the original questions and should reveal if is it possible for grains and cereal crops to be replaced in part or supplemented by tree products in the context of a more sustainable agriculture.

Chapter 5

Trees for Energy

Trees grown especially for energy production purposes, other than firewood, deserve a chapter of their own. Biofuel from tree products is something of an innovative use of tree crops. The redirection of research into this matter now is being spurred by the rising costs of petroleum and other fossil-fuel products, and also by an anticipated difficulty with obtaining adequate future supplies. Research into tree species which can be cultivated and harvested to produce energy has also been spurred by fears that some biofuels, instead of improving the energy situation, will simply take food out of the mouths of the poor as maize, sorghum, and other vital grain crops are turned into substitute fuels. Another fear is increased demand for tropical oil seed crops, instead of improving the environmental equation of energy production, will lead to a further loss of forest cover and natural diversity as new plantations for growing oil-rich seeds and foliage are developed at the expense of forests and jungles.

So, what are biofuels? Biofuels are crop products that can be processed into liquid fuels for the purpose of transportation, on-farm energy, or heating. Bioethanol and biodiesel are the most common biofuels, but there are others, such as biomethanol, biogas and biomethylester. Production of these latter three biofuels is still in the experimental stage, though many small farms use biogas and even biomethanol. Not many areas are planning to use these on a large enough scale to replace fossil fuels.

Bioethanol

Bioethanol currently looks like the most likely substitute for gasoline. However, there will be a long process of development, evaluation, and experimentation before bioethanol can be produced sustainably. Bioethanol is now commonly produced from maize, beets, wheat, sorghum and sugar cane. These crops are all important food crops. The use of them to produce fuel drives up the price of these crops world wide. This is a major problem for the poorest people who grow or purchase the crops for food.

All these crops require considerable inputs in fresh water energy and labor causing many researchers to wonder if the apparently small net energy gain is truly worth the loss in food and potable water. Other researchers have declared turning maize and other food crops into ethanol is a process which on the surface requires 29% percent more fossil-fuel energy to produce than it yields when used—and if all the costs and subsidies are added up, including repairing damage done to waterways and soil by industrial style farming, corn-based ethanol costs a stunning seven times as much to produce as it yields as a biofuel.

The creation of ethanol and cellulosic ethanol are processes which illustrate both the potential and the problems of biofuel production. Conventionally, ethanol is currently derived from corn, wheat, or soybeans, but corn is the most used feedstock. In dry milling operations, liquefied corn starch is produced by heating corn meal with water and enzymes. A second enzyme converts the liquid starch to sugars which are fermented by yeast into ethanol and CO_2. Wet milling operations separate out oil, fiber and protein from the starch before it is fermented into ethanol but the results are the same. Making ethanol in these manners utilizes fossil-fuels during the conversion processes, drives up the price of the food crops being used for the raw materials, and ends up putting the poorer citizens of the world in direct competition with the automobile.

Many people, consumers and researchers as well, are understandably revolted by the idea of burning food for fuel. This fact alone should keep corn, wheat, and soybeans from being used to make bioethanol. Unfortunately, this kind of competition for resources is inherently unfair and a full gas tank will win out over a full belly every time unless concerned citizens push for legislation which will eliminate subsidies and keep staple food crops from being used for fuel. Corn, wheat, and soybeans are actually far too valuable to use in this manner. They only appear to be cheap raw materials because their price is kept artificially low by the systems of subsidies in the countries where they are produced. The entire concept would be abandoned if the real price of these staples was paid by those who want to convert grain and foodstuffs into fuel.

Often, unwise practices in agriculture are condoned because there is some kind of "economic" benefit and the logic of the "bottom line" is used as an excuse for many abuses. When looking closely at the economic scorecard for corn-based ethanol, one can only wonder what sort of convoluted reasoning and "cooked books" can give it the name of a "green fuel".

Still, ethanol itself holds promise because it is a fuel which burns cleaner than gasoline and it can be made from any material containing cellulose. It does *not* have to be made by letting food crops go up in smoke.

Biodiesel

Biodiesel is currently made from seeds such as rapeseed, sunflowers, soy, palm, coconut, and jatropha.

With the exception of jatropha seeds, all these crops are also edible or used to make edible oil. Except for jatropha, they currently are massively cultivated mono-crops, ruinous to the countryside and harvested with energy intensive methods.

The palm oil plantations are especially devastating, causing the loss of much of the remaining Asian rainforests and large tracts of the South American jungles as vast areas are cleared to make way for the planting of oil palms. This is obviously a costly and unsustainable business as well and can only be advertised as a “renewable” and “ecological” way of making fuel substitutes if there is a willing suspension of disbelief on the part of everyone involved.

Can Biofuels Be Produced Sustainably?

Instead of asking, “Can biofuels replace fossil fuels?” perhaps the better question would be, “Can biofuels be produced sustainably for a more sustainable economy?” If biofuels could be produced sustainably, the very fact they are something which can be grown, rather than a one-time, non-renewable endowment, would change the patterns of energy production and use in agriculture.

While it is not possible to power up the current gigantic and wasteful global systems of cultivation, harvesting, and transportation by switching from gasoline to bioethanol, it may be possible to produce from a given locale's agricultural waste and some deliberately cultivated energy crops, enough biofuels to take care of the energy needs of that specific location. This would mitigate shortages of fossil fuels and return energy autonomy and energy independence to many agricultural communities. Local production of biofuels would also save the energy being lost in the transport of fuels produced centrally. Local biofuel production will also create local jobs and job opportunities. Such a strategy would be an important step in the relocalization of agriculture.

However, this strategy would require the identification and usage of appropriate crops for each area which are suitable for biofuel production. Certainly a healthy stable of local energy crops would include trees, but probably it would include perennial grasses and other copiously growing plants which can also be used.

Sorghum, for instance, is an interesting biofuel feedstock which would be appropriate for many areas. Sorghum needs only a fourth of the amount of water and energy which is needed to cultivate corn and can produce 4,000 to 6,000 gallons of conventional ethanol per acre per year. Sorghum has a high yield of biomass, a high percentage of fermentable sugars, a high percentage of combustible material for fueling the process of converting it to ethanol, and a comparatively short growth period. It is also in demand for bread, beer, and animal feed but sorghum is so hardy and so diverse it can be cultivated in degraded and marginal lands where corn, wheat, and soybeans cannot. Increasing the amounts and types of sorghum under cultivation may be one sound way of increasing conventional ethanol production capacity without depriving the poor of food. Perennial sorghums can be used for topsoil regeneration as well as producing biomass for fuel.

Cellulosic vs. Conventional Ethanol

Cellulosic ethanol can be made from many forms of agricultural waste, from the byproducts such as sawdust and paper pulp of many industrial processes, and from crops of all sorts which have high amounts of cellulose, including trees grown especially for this purpose. The cellulosic biomass from these many sources which can be used for ethanol feedstock is composed of cellulose, hemi-cellulose, and lignin with small amounts of lipids, protein, and ash also present. Roughly two thirds of the mass is cellulose and hemi-cellulose and roughly one third is lignin. Fermentable sugars are extracted from the stock by breaking down polysaccharides through acid hydrolysis or enzymatic hydrolysis. After the sugars have been extracted from the biomass, microbial fermentation of this sugar yields ethanol and CO_2.

Lignin, a by product of hydrolysis, is a fuel with the energy value of coal which burns with almost no greenhouse emissions. This allows the process of making cellulosic ethanol to be fueled by one of its own by-products and totally eliminates the need for fossil-fuel input in the ethanol conversion process.

Additionally, cellulosic ethanol will not take food out of the mouths of the poor (and put it in the gas tank of vehicles) because it can be made from such a wide variety of feedstock, most of which are not in the human food chain. It can also be produced from materials which are usually burned or put into land fills, such as paper sludge and municipal garbage. Its production can be accomplished by using a by-product of its own processing for energy to covert biomass into ethanol. Because it can be made from so many materials, it can be produced anywhere there is agriculture, trees, or a human population.

While many fear genetically engineered trees may be planted to produce ethanol or indigenous people and small farmers may be pushed off their land because of the demand for bio-energy mono-crops, cellulosic ethanol can be made out of so many waste products, perhaps 40 to 50% of potential global demand can be produced without planting anything new at all. The remaining requirement can be gained by planting a wide variety of high cellulose crops specific to areas where biofuels are needed in an ecologically friendly manner. These crops should be regenerative, planted on degraded or worked out land, non GMO, water and labor resource thrifty, and genetically diverse. In this way, every area will be encouraged to produce its own ethanol and little or no energy will be lost in trucking it about.

Arboreal Energy Crops

The ever useful mesquite tree, used for food and livestock feed, is also a potential bio-energy tree. One ton of dense mesquite wood will yield 200 gallons of ethanol. An acre of mesquite standing 3 to 4 meters tall will yield 8 to 10 tons of wood. Above ground growth can be harvested and then the stumps can be allowed to regenerate for reharvesting in a few years. Mesquite grows well in areas with more than 300mm of rainfall and is both salt tolerant and drought tolerant.

The poplar tree is also a serious energy crop candidate. It is often selected for its swift growth and its adaptation to many different climatic zones, as well as, its relatively long growing season. Poplars can be harvested by coppicing at a rate of 10 tons per acre per year to be made into 700 gallons of cellulosic ethanol. The residue from ethanol production can be composted and returned to fertilize the soil where the trees are grown. Poplars and other swift growing hardwoods can be harvested sustainably for decades.

Some researchers have discovered the potential of these trees. Experiments to change the lignin content and increase the amount of cellulose and hemicellulose in the trees are underway to facilitate ethanol production since the presence of lignin makes extraction more difficult. However, the GM redesign of the poplar may be dangerous from an environmental point of view and unnecessary from an economic one. Lignin is a high energy content substance which can replace fossil-fuel input in ethanol production. Since it is such a useful substance, a GM redesign to reduce the percentage in poplar wood makes little sense. Trees with reduced lignin may be less resistant to weather events and more prone to breaking, bending, and falling.

Cottonwood trees also grow swiftly, up to 3 meters a year, can be coppiced and also have high energy potential. They can be harvested at rates of 9 to 12 tons per acre per year, producing from 650 to 950 gallons of ethanol per acre per year. Hybrids of poplar and cottonwood combine swift growth with high biomass yields and dense wood. Other trees being investigated as possible energy crops are willows, albitzias, wattles, horse chestnut trees (the "conkers" are also being considered as a source of biodiesel), leucaenas, marulas, and tamarisks. There are doubtless dozens of other plants and trees which could be cultivated as energy crops. they simply have not been investigated or evaluated as yet.

Sustainable cultivation practices would put cellulosic ethanol at the head of the pack of biofuels, making it not only the least expensive of the possibilities but also the "greenest." Biodiesel could then be produced without exacerbating the current problems of agriculture. About 10% of the current biodiesel feedstocks are used vegetable oil and waste vegetable oil. This means the other 90% must come from oil rich crops. The oil palm cultivated as a mono-culture and doing untold amounts of damage to communities, watersheds, forests, and local plant associations all over S. America and tropical Asia is not the most efficient of the oil producing plants.

Wild Trees for Energy

Ximena caffra, Moringa olifera and Pappea capensis are wild African trees with oil rich seeds which can be converted to biodiesel. With each tree producing an average of 10 kilos of seeds and 65% of the oil being convertible to biodiesel, a hectare of trees will produce 2,400 liters of oil or 1,560 liters of biodiesel each year. Selection and cultivation of the trees would probably increase the yield but thousands of hectares of these trees are already present in the "bush" of southern Africa. The seeds can be gathered presently by wild croppers, children, and the unemployed without damaging the tree populations and then used for local energy needs.

The most efficient oil or ethanol producers for some areas may actually be in the families of shrubs and the perennial grasses, renewable plants, and regenerative sods which produce biomass year after year suitable for biodiesel and bioethanol production. Native wild plants and eco-typical cultivars are often overlooked as potential energy crops. These should be surveyed and evaluated before plants from other areas are introduced.

Other Energy Crop Candidates

Hemp is a high yielding multipurpose fuel and fiber crop. Hemp can produce four times as much biomass as the biomass of a sustainably harvested forest of comparative area. An acre of hemp yields 10 tons of biomass in four months, enough to make 1,000 gallons of biomethanol plus 300 pounds of oil. Confusion between hemp and its close botanical relative the marijuana plant, have kept hemp from being cultivated in many places, most notably North America. Since hemp contains almost none of the mind altering substances of marijuana and confers so many benefits, there is no reason why hemp should not assume an honored place in the sustainable agricultural systems of the future. Cloth, edible oil, and fuel are only a few of hemp's possible products.

Jatropha curcas, a large shrub which produces oil rich seeds, is being grown along railway lines by India's village cooperatives, utilizing land which was out of production and converting the seeds to thousands of liters of biodiesel which are used locally to power tractors and other agricultural machinery.

Switchgrass, an anti-erosive perennial forage crop, already planted in some areas to stop the loss of topsoil to wind and water, also has great potential for biomass which can be made into fuel, producing six tons of dry matter per acre annually. Standard breeding techniques could double this amount, making switchgrass a leading crop candidate. The perennial grass, native to North America, would be mowed twice a year in moderate climates and once a year in cold ones, and planted in unstable, eroded, and degraded areas which are currently out of production. The environmental benefits of this crop are many. The foremost benefit, after biofuel production, is the stabilization of eroded and damaged land, pulp by-products for animal feeding, and increased water absorption by the switchgrass swales and sods.

There is no lack of agricultural materials to make sustainable bioethanol, nor is there a lack of the know how to take the materials and convert them into fuel. What is missing is the commitment to retooling the systems which use energy so less energy can be used to accomplish the same important tasks on a smaller, smarter, greener, more flexible scale.

A radical reorganization of farming systems must come sooner or later if agriculture is going to continue successfully in the developed world. As well as finding alternative sources of energy, alternative ways to use energy must be found. Energy crops must be diverse and keyed into the specific locales where they will be used in conjunction with the locale's agricultural, and perhaps municipal waste as well. To be truly efficient, the local biofuels must be utilized close to the source rather than produced centrally and distributed over long distances.

Since one of the advantages of biodiesel and bioethanol is its transportability, it could well be asked, "Why use energy to convert mesquite to ethanol? If the maximum net energy is extracted from materials by using the materials as close as possible to the source, why not just use the mesquite wood itself as fuel?" If we want an answer to this question and others which come to mind, it requires new ways of thinking about the issues.

Inventors and researchers are currently designing tractors that use wood as fuel and automobiles which use manure. It is already clear these machines will be smaller, lighter and more energy efficient. They will have to be to utilize less energy dense fuels than fossil-fuels.

There is not going to be a neat, organized and seamless transition from fossil fuel to biofuel use. Biofuels are not a substitute for oil in quantity or versatility. If made and used wisely, taking care their production is done in such a way the conversion is accomplished close to their usage, all the hidden costs of acquiring the feedstock are taken into consideration, and if there is a net energy gain in the process, it may be possible to produce enough energy to keep the most important processes of food production up and running. Biofuels cannot power up the millions of automobiles on the roads, nor can they fuel the fleets of planes in the sky.

There are going to be many flat tires, bumps, detours and mistakes along the way to the sustainable production and use of biofuels, but to reach the desired destination, researchers, inventors, and consumers should be able to agree on a few simple premises. At a minimum, these are:

1. Bioethanol will not be made from staple food crops
2. There will be no subsidies for crops turned into bioethanol or biodiesel
3. Bioethanol and biodiesel will be produced locally and used in scaled back energy efficient systems.
4. The current systems of agriculture and the transportation of food will be redesigned to use much less energy, much more efficiently.

In this way, bioethanol and other biofuels can come into widespread use without adding to the problems of agricultural systems which already make them problematic and unstable. Trees and perennial plants will be planted and cultivated sustainably for the production of energy which is both renewable and "green."

Chapter 6
Fuel and Firewood

One of the continual challenges for the poor is finding enough firewood for fuel to cook their meals. Millions cook over open fires or in small wood-fired stoves and ovens. Since the grains, pulses, and tubers which make up the bulk of their diets have to be cooked to be edible, there is fierce competition, what the National Academy of Science calls a "frantic scramble" and the FAO calls the "cooking pot wars," over anything which can burn.

In very poor countries up to 90% of the population prepares food in this manner and the head of a family can use over a ton of wood a year. Because of this, firewood is becoming harder to find. The developing world is in the midst of a critical shortage of this essential material. Prices for firewood and fuel have soared. More time must be spent to acquire wood for fuel. Often, people who cannot find it or buy it are forced to burn animal dung or crop residues, thus depriving the soil of these two vital inputs and causing decreases in crop yield. In many places it takes more time to find fuel to cook the food than is expended in growing the food to eat. To quote the National Academy of Sciences (NAS) publication *The Lost Crops of Africa, vol. 1* "There is a saying in Africa that it costs more to heat the pot than to fill it."

Obviously this need is causing many forests and woodlands to shrink and some to even disappear altogether, as the speed with which the wood is harvested and used is greater than its rate of renewal. The result is often a treeless, eroded, dying landscape and a population which can no longer feed itself or even cook the donated grains of food aid programs.

In recent years, research into selecting species which might be cultivated to help provide this vital fuel resource has led to accelerated tree plantings at the level of the village and the family, many reforestation projects, and the establishment of fuel tree plantations in some locales. While population growth in the poorest countries has tended to swallow up the modest gains from some projects, other efforts to plant and maintain firewood plantations and to use them sustainably have been very successful. The most successful efforts by governments or NGOs are those commanding broad support in nearby communities.

Planting millions of trees, nurturing them to maturity, and keeping them from being poached by desperate people, eaten by hungry domestic animals, or harvested unsustainably is complicated and difficult under the best conditions. In areas where there have been inadequate supplies of wood for a very long time, it is very difficult to keep even the fastest growing trees from premature harvest without a local consensus in an agreement that not harvesting the fuel and wood today means there will be more fuel and wood for tomorrow. To reach this critical agreement requires support and sanction from the traditional power structure, investment by national government or a powerful NGO, approval of the local government, and the cooperation of a good percentage of the local population. Then it may be several years between the planting of the firewood plantation and the first harvest. For people who do not have wood with which to cook, there is a terrible temptation to use the resource as soon is there is anything big enough to burn.

Short Term Fuel Crops

Luckily, there are some plants which produce usable biomass for fuel within a few months. If a plantation of this sort is established in the same general area as the firewood plantation, there is a much greater chance of preserving the plantation trees until wood can be harvested. Some short-term fuel plants also produce edible products and these should be considered for fast producing fuel plantations in addition to quick growing trees.

Solid stem sorghum is one such plant. It has a full, small head of grain but a solid, fibrous, almost woody, stem which is easy to set alight and which gives off much heat as it burns. One type from Egypt, called Giza 114, has a particularly lignified stem and has been used in Egypt and Peru for cooking, firing brick kilns, and baking ovens.

Sorghum is a fuel crop which can be harvested in three to fourth months, with three crops grown per year in areas with an appropriate climate. The annual sorghum biomass equals or betters the amount of usable biomass which could be gained from most tree species, topping 75 tons per hectare. When dried and the grain removed, the stalk and root mass of a hectare would equal more than 10 tons of wood. Unlike bulky firewood which must be chopped up to be used, the sorghum stalks fit nicely under a pot propped up on stones or bricks and can be broken by hand and fed into the small cook stoves used in most poor rural areas.

Another possible food/fuel crop is corn. The actual corn kernels are too valuable to be used as fuel, but the cobs and stalks burn well in cook stoves. Corn stalks have many other uses and do not produce as much heat as solid stem sorghum, so using corn should be considered something of a stopgap measure.

The pigeon pea (**Cajanus cajan**) is a food crop with tall woody stems which burn as well as small branches do and it gives off more heat than kindling. The nutritious seeds of the pigeon pea contain 22 percent protein. Green seeds and fresh pods are often eaten as vegetables. The pods husks and foliage are good fodder for animals and silkworms have been cultivated on pigeon pea leaves. The pigeon pea grows quickly and densely enough to be used as a windbreak in sandy areas. The plant is particularly well suited to arid, saline, and infertile areas. Four hundred mm of rainfall is enough to guarantee a bountiful crop and the plant tolerates a wide range of sandy soils and loams. Besides the edible seeds and nitrogen fixing ability of this hardy legume, in four to nine months the plants reach maturity and 2-3 tons of fuel can be harvested per hectare of the brittle, round, woody, lower stalks while the thinner upper portions can be used for thatch or basketry.

Among the inedible plants there are many short-term fuel crop candidates. Reeds, elephant grass, Sudan grass and the smaller bamboos burn fiercely enough to fire up any stove or oven. They also grow quickly and re-grow from the remaining roots after being cut or mowed.

Having a short-term fuel crop close by often stops the desperate foraging in whatever woodlands and groves still remain and allows firewood plantations to reach maturity when sustainable harvesting of wood can begin.

Fast Growing Woodlot Trees

There are many fast growing tree species which can be used for both firewood and fuel. Fuel and firewood species including trees, shrubs, and other kinds of plants, are usually multipurpose species which are easy to establish, adapt well to different sites, and need little care. They are often planted on marginal lands, or areas which have degraded. Often they are species which are not attractive as browse for goats and sheep or food for wildlife.

Many of these plants can be coppiced or re-harvested, have nitrogen fixing ability, and burn without much toxicity, sparks or smoke. Some of these plants grow in a wide variety of environments, needing different amounts of rainfall, intensity of sunlight and different kinds of terrain.

Because of the danger of invasiveness in the kind of quick growing, hardy plants chosen for firewood and fuel, some effort had been made to give local trees a higher priority. Luckily, there is no lack of candidate species for fuel and firewood plantations.

Focusing on Fuelwood Species for Arid Areas

In semi-arid and dry areas, the shortage of firewood is more acute simply because the aridity severely limits the kind of plants which will grow there. The natural biomass productivity of these plants is usually low due to the lack of rainfall. Because of these limitations, this chapter deals with the kind of hardy plants which can alleviate this shortage.

There are many more species suitable for fuelwood in temperate and tropical humid zones than ones which grow in arid and semiarid zones. A list of these species and their description can be found in the National Academy of Science (NAS) booklet *Firewood Crops – Shrub and Tree Species for Energy Production*. An abbreviated list can be found in Appendix I of this book. For the arid, semi-arid, and saline areas, there are some suitable trees, described in the latter parts of this chapter along with some details as to their tolerances, as well as their positive and negative qualities.

Many millions of people live in arid and semi-arid areas, usually defined as areas with less than 500 mm rainfall and many of them are desperately poor. In their need for firewood and fuel they damage fragile areas, removing trees, uprooting saplings, burning anything combustible, and then burning the dung of their animals which contains nutrients which should have been returned to the soil. Without trees to bind and shade the soil and to allow for the infiltration of rainwater, grass and other vegetation thins out and vanishes. Aquifers do not recharge. Grazing is poor and hungry animals begin to eat leaves, twigs, and saplings so the area does not recover its tree cover. The result in many cases is man-made desertification, a state of ecosystem collapse in which virtually all vegetation is gone and even the soil blows away on the wind.

Trees for firewood plantations in these areas are usually tough, salt tolerant species with especially deep or well-developed root systems which penetrate to subsoil moisture or win moisture from extremely large areas. Often they are trees which are associated with nitrogen fixing bacteria. Most have small leaves or needle-like leaves to limit transpiration or some other physiological mechanism to limit water loss during drought. Many have thorns to protect them from hungry animals.

Select Individual Species of Useful Trees and Plants

Adhatoda vasica, is a many-branched evergreen shrub which usually grows between 2 and 6 meters in height. This plant is almost never grazed by sheep, goats or wild animals. Because of this, it can be planted as a hedge, a windbreak, or a living fence. It is a lush looking, leafy plant with broad leaves which give off an unpleasant smell when crushed. The white and purple flowers are also unpleasantly scented. The wood, however, burns cleanly and brightly, giving off considerable lights and heat. Native to India, Burma, Sri Lanka, and Malaysia, it is often turned into charcoal. It is also a medicinal plant. Its leaves are used as green manure or to curtail insect populations much as Neem leaves are. The plant is also a source of yellow dye. Its usefulness is restricted to tropical and subtropical areas because it is sensitive to low temperatures and frost.

Albizia lebbeck, is a robust and adaptable tree with a full crown and white bark which produces small timber, fuelwood, fixes nitrogen through nodules in its roots, and is excellent for reforesting dry, alkaline, sandy areas. The tree can reach 30 meters in height under the right conditions, but is usually between 6 and 15 meters tall. The wood of the tree is dense and the tree coppices well. Leaves and pods provide good fodder or green manure. The tree is valued for shade, for beauty, and for its value to bee keepers since the tree flowers copiously early in the spring. Albizia lebbeck trees are also considered suitable for erosion control both in windy desert areas and areas where light soil is in danger from water erosion. It is somewhat frost and drought tolerant. In its native range, the tree does well from sea level to 1600 meters with rainfall ranging from a modest 40 mm to a “wet” 200 mm.

Anogeissus latifolia, is sometimes called the Axelwood tree. It is a rugged tree of the semi-arid zones of India which produces fine, dense firewood. The trees develop into large plants in areas of deep soil, but stay gnarled and stunted in areas where the soil is rocky. The trees coppice well and are fire hardy, but will grow slowly in very dry conditions. The gum of the tree (Gatty gum) is twice as viscous as gum Arabic and is used in medicine and printing. The bark and leaves are rich in tannin which is used in the leather-making industry of India. The foliage is suitable fodder for grazing animals and can be fed to tasar silkworms. The tree's name, Axelwood tree, comes from the strength and durability of its wood which polishes well and can be used to make farm implements, including the shafts and axels of carts. Currently this non-domesticated species is propagated by seed.

Azadirachta indica, or Neem tree is also known as the "village pharmacy tree." Neem is potentially one of the most important and valuable of the arid zone trees. It can grow in arid and nutrient deficient soil and be used as a fast developing source of fuelwood. It can also be cultivated as a medicinal plant, since virtually every part of the tree has value in both Ayurvedic and conventional medicine.

Neem is a deep rooted, medium size tree with beautifully feathery deep green foliage and small fragrant white flowers. The trees generally have a short trunk or multiple trunks covered with thick reddish grey bark. The wood of the neem is reddish white and durable. The wood is heavy and makes suitable poles and other small timber items. It is rarely attacked by insects and polishes well. After the first season, the development of young plants is rapid. Though cold tolerance in adult tree varies from tree to tree, young neem trees are all very vulnerable to frost and low temperatures. The tree coppices easily and regrowth develops even faster than the original trunks.

Besides the neem wood, the seeds contain oil suitable for lamps or to lubricate machinery. Neem oil is also a wound healer and disinfectant. Neem is an excellent windbreak and shade tree. Its heavy leaf falls create natural mulch and the tree has been used in many reclamation plantings in Africa and Asia. Neem bark is full of tannins useful in the leather making industry.

Neem products, which are also fungicidal, are added to toothpaste, shampoo and soap. Several "green" preparations for insect control are made from Neem products. Small neem twigs are used as chew sticks and toothbrushes because of the beneficial effects of Neem sap on teeth and gums.

Neem trees can be grown from cuttings but they are most easily grown from fresh seed. Seed viability is severely limited. Most seeds will lose their ability to sprout after a month. The plant is extremely heat tolerant and does better in saline soils than might be expected if enough water is supplied and drainage is good. The tree can survive with as little as 130 mm of natural rainfall or as much as 500 mm.

Cajanus cajan, or The pigeon pea, mentioned at the beginning of the chapter, is not a tree but a food crop with tall woody stalks. These stalks are an important by- product because they can be used instead of wood to cook meals in rural households. A nitrogen fixing legume, the pigeon pea can be grown in infertile or worked out soils. The speed of plant development protects soils quickly from erosion and sometimes a double row of pigeon pea plants, which may be 3.5 meters tall is used as a living fence or a windbreak.

Dry seeds of the pigeon pea are the plants most important product, but green pods are sometimes harvested and used as a vegetable. The pods and foliage are suitable forage for grazing animals and potential food for silkworms and lac insects, the tiny creatures from which shellac is obtained.

The plants grow in full sunlight in hot areas with rainfall over 500 mm and below 1000 mm. Pigeon peas are not frost tolerant. They will grow in a wide range of poor soils but best in deep sandy loams. The plant is propagated by sowing seed “in situ”.

Cassia siamea, or Yellow cassia, is a large evergreen tree with dense dark colored wood which makes a good fuel or an attractive timber for cabinet making. The trees coppice easily and are hardy enough to be chosen for land reclamation and reforestation projects in many countries. In Nigeria these trees are used successfully to replant and reclaim mining sites.

The tree can be propagated by seed and is tolerant of heat as long as it is generously supplied with water for part of the year. So the areas most suitable for this tree are monsoonal areas with a long dry season, canals, or riverbank sites. The tree is not cold tolerant nor will it grow in areas with less than 500 mm of rain per year. The leaves and pods are also palatable for sheep, goats, and cattle, so young trees should be protected for the first years. Pigs should not be fed any part of this tree, as the leaves, pods and twigs are toxic to these animals.

Colophospermum mopane produces the best firewood in Africa. It is also the host of the African delicacy, the Mopane worm. A moderately tall tree, with fissured gray bark, the Mopane grows on very poor and damaged soil where little else can grow. Mopane leaves have a distinctive butterfly shape and the wood is so hard, felling a tree is a major undertaking. The durable wood is useful in construction for beams, fence posts, poles, bridge supports, and railroad ties. The leaves provide excellent fodder for sheep, goats, horses, and cattle and their percentage in the food ration of grazing animals does not have to be limited. The leaves retain their food value when dried. The Mopane worm, actually a caterpillar, grows and develops on a diet of Mopane leaves in great numbers. These worms are collected and eaten dried, fried in oil, or salted and are considered a delicious snack. The tree is very salt tolerant and will grow in shallow soil but does best in deep loam. The tree is heat tolerant to a high degree and can tolerate mild frosts but heavy frost can damage the branches and cause dieback.

Emblica officinalis, a semi-domesticated tree, is very much appreciated in Asia and is usually left standing when land is cleared for agriculture. A medium sized deciduous tree with slender branches and pale green or reddish fruits, the Emblica is used for reforestation and land reclamation, cultivated for firewood, or for its extremely close grained and durable timber. It is also planted in gardens for the sake of its fruit, which is rich in vitamins and pectin.

The Emblica wood is water resistant and used for building in wet areas and for lining wells. The fruit is picked, dried, or cooked, and appears in candied forms and as relishes and chutneys. Both fruits and leaves are eaten by livestock and the fallen leaves have considerable value as green manure. The trees are frost sensitive and do best where there are two distinct seasons. The tree does not grow well above 600 meters in altitude. Native to humid areas, it also grows in dry zones and tolerates sandy and alkaline soil. Seedling Emblicas are inferior to budded, more domesticated types. The seeds are not viable for more than a few weeks and germinate well after being treated with hot water.

Parkinsonia aculeata, also called Jerusalem Thorn or Blue Palo Verde, is native to the US southwest but naturalized in Hawaii, South Africa, and India. It has also been introduced to Israel, Jamaica, Florida, Cyprus, Uganda, Kenya, and Chile. It is a small, crooked tree which rarely reaches above 8 meters in height. It has a green trunk, about 40 cm in diameter, many drooping branches, and delicate grass-like foliage.

Despite its tender look, Jerusalem Thorn is a very hardy tree. It is most comfortable in full sunlight and poor, gravelly or sandy soils. It is heat tolerant into the 40° C range and tolerates light frosts. The tree is appropriate for anti-erosive plantings but also prized for its vivid foliage and flowers. These qualities make it a candidate for urban plantings, as well as, planting along roads and highways threatened by sand encroachment. The tree also fixes nitrogen and is tolerant of saline sites. A minimum of 200 mm of rainfall is enough for reasonable rates of growth and development, but waterlogged soils retard growth and can kill young trees. Because of this, drainage must be good where the Jerusalem Thorn is planted even if the water supply is sparse or seasonable. The wood is used for firewood and charcoal in Mexico and other South American countries but is brittle and not suitable for lumber. Trees revive easily even after vigorous harvesting. Branches are lopped off to feed goats and sheep, and the seed pods are very attractive to grazing animals.

Pinus halepensis, the Aleppo Pine, is a tree suitable for Mediterranean sites with 250 to 800mm of rainfall and 7 to 8 rainless months. It is a tree with a round crown which may reach 25 meters in height on deep soils, but it is able to tolerate poor, shallow and eroded areas. It is not suitable for swampy soils but grows well on clay soils which many other conifers will not tolerate. The tree is tolerant of cold down to minus 20° C and heat into the 30° C range. It can grow up to 2,000 meters in altitude.

Aleppo Pine is a useful source of tinder and resin. It is a robust tree for soil conservation plantings and shelter belts, and also a popular ornamental. It has been planted extensively in Greece, Italy, Israel, Jordan and many other semi-arid areas around the Mediterranean for reforestation projects. It has also been introduced to South America and the Soviet Union, especially to the warm semi-arid areas there. In some areas it is considered too valuable to use as firewood and the trees are selectively harvested for timber. The wood is resinous and burns well so waste wood from logging for construction wood is collected and sold as well. The trees are planted out as seedlings, having been sown in pots and beds in the spring and then moved into plantation format or terraces at the start of the rainy season.

The Acacia Genus of Trees

The Acacia genus is a large family of trees particularly suited to arid, saline, and infertile areas. Among the many types of Acacia, several species stand out as particularly good trees for fuel and firewood.

Acacia brachystachya, also called the umbrella mulga, turpentine mulga or false bowgada. This is actually a large bush which grows up to 7 meters tall with multiple trunks. It grows in the Australian interior in very arid conditions, mostly in areas where the temperature does not drop below freezing but may reach as high as 58° C. The wood of the tree is hard and heavy and the foliage can be eaten by stock. Seed production is irregular, but the seeds are edible. Some related species of similar potential are:

Acacia mangium, a very fast growing species from Australia's tropical edge, which is also a good timber tree and can reach a height of 15 meters in three years,

Acacia lysiophloia, also a large shrub from areas with very poor soil, the wood burns hotly, but the sticky leaves make unpalatable forage and,

Acacia holoserica, a small tree which needs more abundant rainfall, at least 500 mm, but has abundant seeds and matures quickly.

Acacia cambagei, also know as the Gidgee or the Stinking Wattle. This is a very tall, open habited acacia with pungent gray-green foliage. It is found in western Queensland and New South Wales. It has an extremely heavy wood which burns so hotly green or dry that it is usually mixed with less flammable materials. The wood is also known for its dense grain and resistance to termites. The areas in which this tree thrives have relatively low rainfall but the temperatures rarely rise above 35° C and the tree seems to grow well in hilly areas.

Acacia cyclops, this tree is extremely tolerant of salt, salinization and salt spray. It is a dense evergreen multiple stemmed shrub which spreads out laterally and forms and dense mat of roots. Because of this habit, the tree is often planted to stabilize coastal areas or dunes. While the tree is native to Australia, it has been used in many other countries for coastal planting and land reclamation projects. While the logs rarely exceed 20 cm in diameter, the wood burns green or dry and is considered a superior wood for cook stoves and kilns. The foliage is browsed by stock and wildlife and the seeds are often eaten by these animals and dispersed by birds. The tree survives in areas with as little as 200 mm rainfall, but has high light requirements and while tolerant of slight frosts, it is unhappy in areas with regular temperatures above 40° C.

Acacia nilotica, also called the Egyptian thorn or the balbul, is a very valuable source of timber, fuel, fodder, tannin and honey. It can grow to a height of 20 meters in good conditions, but its size is usually limited by available water. In arid areas, it may not exceed six meters in height and in extremely dry zones it may resemble a shrub. Acacia nilotica is distinguished from other acacias by its extremely fragrant, bright yellow flowers and the paired white spines at the base of each leaf. The trees are also very thorny.

It is believed Acacia nilotica was the earliest commercial source of gum Arabic. The ancient Egyptians used the tree for boatbuilding. The wood is very heavy, resistant to termites and suitable for carving, making bowls, plates, cutting boards and eating utensils. The tree is native to Africa and found in semi-arid and arid zones from Egypt to Botswana. It grows in a variety of soils, from alluvial to heavy clays, and develops considerable resistance to frost as an adult.

A subspecies Acacia nilotica indica is propagated by direct seeding and used extensively for fuel, building, and grazing in Pakistan and India. This is a very thorny and difficult to handle tree which should only be introduced to areas critically in need of firewood or reclamation.

Acacia salinga, this is an extremely adaptable and rugged tree which has naturalized itself in some very desolate areas. A dense busy shrub or small tree, its drooping branches are graced with golden flowers and bluish smooth leaves. The tree's full height is reached in five to six years. In areas with little rainfall, the tree will stay small unless its aggressive root system gets into the ground water. While its wood is "sappy," light, and burns quickly, the tree has other uses. It is an excellent dune fixer, windbreak, and fodder tree. Its gum, exuded freely from damaged branches is useful in pickling. The tree tolerates brief periods below freezing. It has a remarkable ability to compete with weeds, however it can be invasive and hard to eradicate.

Acacia senegal, this tree is the source of gum Arabic. It is not a fast growing tree, but it is so hardy it is planted in land reclamation projects in the most challenging desert areas. While the tree is usually small and thorny, in the right conditions, it can reach a height of 13 meters. The tree lives for 25 to 30 years and tolerates both browsing from wild and domestic animals and coppicing by human populations. The wood of the tree is hard and heavy. This makes it excellent for firewood, but also good for poles, carving, and in the manufacture of agricultural implements. Its root fiber is woven into fishnets or twisted into rope. The tree's most valuable product is gum Arabic, used in making medicines, confections, baking, arts supplies and electronics. The seeds and pods of the tree are suitable for human consumption, the twigs and leaves are a good source of fodder and like all acacias, Acacia senegal is a nitrogen fixing plant, excellent for stabilizing sandy soil and dunes.

Acacia seyal, is a resilient, drought tolerant tree of the dry areas of Africa, which is particularly resistant to grass fires and forest fires because of its thick, greenish yellow bark. With the bark removed, the cured wood of this tree is prized as some of the best fuelwood and smokewood in Africa. The tree can reach 12 meters in height, but it is usually much smaller, thorny, and flat-topped, often with multiple trunks. Found in dry savannahs, it is also an important fodder plant. The exuded gum is also valuable, darker in color and generally inferior to gum Arabic, but with many of gum Arabic's qualities.

Acacia tortillas, this tree grows in some of the driest areas on the planet. It is a medium sized tree, usually from 4 to 10 meters tall, although trees in some areas have reached 15 meters in height. In very arid zones, the tree remains small; its branches tangled and gnarled into a flat, thorny, canopy. The dense red heartwood makes excellent charcoal. Relative to other arid zone plants, the Acacia tortillas is relatively fast growing. The pods of the tree are a favorite food of broth wild grazers and domestic animals. Grazing animals also eat the leaves and twigs of the tree. The tree favors alkaline soils and can survive in areas with less than 100 mm rainfall and long dry seasons. The wood is also used for making fence posts and small agricultural implements.

There are many other species of Acacia trees which are useful for fuelwood, grazing, and carving, but the preceding Acacia species are the more common and easily found of the Acacias.

Eucalyptus Genus of Trees

Eucalyptus is another genus of trees with many useful members, including trees for wet, temperate, and arid zones. The best types for arid and semi-arid zones are listed below.

Eucalyptus camaldulensis, is a tall, slender stemmed, tree with cream colored or pinkish smooth bark. The leaves are long, tough and have a bluish cast. The tree is tolerant of dry conditions, salinity, and it is reasonably hardy to heat and cold. It can be found in areas with only 200 mm of rainfall and in areas with over 1,000 mm of rain. This tree will grow in a wide variety of soils, the leaves are not eaten by livestock or wild animals so in most places where the trees are planted, they grow quickly and without interference by animal life. When young these trees do not compete well with weeds and the planting should be weeded until the tree gains strength and height. The wood produced by this eucalyptus type is an excellent fuel and valued as firewood, and for making charcoal. The tree itself is often planted as a windbreaker, in shelterbelts, and in farm woodlots.

The tree is very much valued by bee keepers as the honey produced from the flowers has a mild pleasant flavor and a fine gold color. It may be the most widely planted of all the eucalyptus types and has been introduced to the Middle East, the Mediterranean countries, and South America. It is also regarded as the most medicinal of the eucalyptus trees. Its leaves, flowers, seeds, and oil are considered valuable materia medica.

Eucalyptus citriodora, is a tall, fast growing tree with a straight bole, reddish or bluish bark and has a crown of pendulous foliage. It is much valued for timber and may reach heights of 45 m in less than a decade in areas with sufficient rainfall. While hardy to high temperatures and light frosts, it needs more rain than Eucalyptus camaldulensis, a minimum of 600 mm for health, but for optimal growth, approximately 900 mm per rainy season is required. If sufficiently watered, a dry season of 6 to 7 months is well tolerated by this tree. Well drained soil is best for this tree but the tree also tolerates soil with some clay. The tree can be propagated by seeds and competes well with weeds.

The wood of a Eucalyptus citriodora is hard and heavy, with low shrinkage during drying. It is an excellent saw timber and used in construction, to make tool handles, poles, posts, and heavy beams. The lemon scented leaves are pressed for a perfumed oil which is rich in citronella. The abundant flowers yield large quantities of citron flavored honey. The tree is so beautiful it is used in ornamental planting in appropriate areas, often to give a "canopy" to a mixed planting.

Eucalyptus gomphocephala, also known as the Tuart, is a tree particularly suited to sandy soils, especially those high in limestone. It tolerates temperature down to minus 4° C and well into the 30's C.

The minimum rainfall for survival is 300 mm per rainy season but the tree does well with 400 to 600 mm and thrives in areas with 600 mm to 1000mm of rain. Sufficiently watered, a dry season of six months is well tolerated. A tall tree with a large trunk, the Tuart may reach a height of 40 meters and a girth of 2.5 meters. Its trunk may be short or forked. Its leaves are thick and shiny and the bark is fibrous and pale gray in color.

The Tuart tree is salt tolerant to some degree and tolerant of active calcium but can be killed by water logging. It should not be planted in very dry areas, as trees weakened by lack of water may be attacked by insects. Also, the plantations themselves may be susceptible to fire.

The Tuart trees coppice readily and produce a dense grayish yellow wood. Annual yields of 21 to 44 m^3 per hectare (ha) have been obtained on good soil in Morocco but the typical plantation, planted on marginal or depleted lands, usually produces 6 to 10 m^3 per ha. The tree is particularly useful in land reclamation and dune stabilization projects in semi-arid areas and has been planted successfully in Cyprus, Turkey, Greece, Israel, Ethiopia, Brazil, and Uruguay.

Eucalyptus microtheca, also called the Coolibah tree, is a tree from the arid zones of Australia. It produces one of the strongest and hardest timbers in the world and makes a good fuel. The tree flourishes in dry areas because of its salt tolerance, resistance to high temperatures, and its ability to resist drought. It is tolerant of mild frosts. The tree is variable in height. It can be as short as three meters or as tall as 20. It has spreading branches with leathery, distinctly veined leaves. This is one of the hardiest of the eucalyptus family and will grow where no other trees can survive.

Coolibah is cultivated for firewood in many African countries and has been introduced as a shade tree in South America and the Middle East. The tree coppices well and the wood itself is considered a more valuable fuel than charcoal made from the boles, because the charcoal has a high ash content. Coolibah timber is difficult to work but is strong and resistant to termites and decay. It makes good fence posts and poles.

The tree can survive with as little as 200 mm of rain and does well in clay soil, alkaline soils, or silty loams. The tree may be propagated from seed but should only be planted out when the seedlings are at least 40 cm high, not sown directly. The young trees must be weeded since the resulting seedlings are not robust and grow slowly through their first months.

Eucalyptus occidentalis, is tree which can be grown under extremely dry and saline conditions. It is also called the Flat Topped Yate. It has been introduced to exceedingly difficult areas in Israel, California, Iran, Sri Lanka, Morocco, and Algeria. Its growth is relatively slow but it produces fine timber, both heavy and strong, which can be used in construction projects. Used as fuel, the wood burns hot with a steady flame.

Flat Topped Yate tolerates brief periods of temperatures close to 0° C and can tolerate highs of 40° C plus. The tree grows well with natural rainfall between 380 and 700 but can also be found in areas with seasonal flooding and on seasonably flooded salt flats. The tree also tolerates clay, gravel, and poorly drained areas which are highly mineralized. It is competitive with weeds in hot, salty areas, and can be propagated from seed. It has not been grown above 500 meters in altitude in any of the reported projects.

The Haloxylon Genus of Trees

The following two plants from the Haloxylon genus grow in some of the harshest conditions on the planet, in deserts both cold and hot and areas of dry frosts and hyper-aridity.

Haloxylon aphyllum, also called Black Saxual, is a xerophytic, halophytic tree which rarely reaches a height above five meters but is very well adapted to harsh environments. The tree tolerates heat well into the 40° C range and cold to about minus 35° C below 0° C. It can regulate its life processes to meet a variety of challenges found in both cold and hot deserts. The trunk of the tree usually has a large irregular base and many large trunks or branches. The thickened leaf stalks the tree has, in place of true leaves, have an odd jointed appearance. The foliage is used for browse and forage.

The tree has been planted in the Soviet Union and China to halt sand encroachment. Plantations have been set out in areas where stabilization and water conservation are important, but Black Saxual is not a pioneer plant and so is introduced to these areas deliberately. Usually saplings are set out after a year or two of development, just before the rainy season, so they will establish themselves while there is enough available water.

The wood is brittle and not usable as lumber but burns as well as brown coal and makes good charcoal.

Haloxylon persicum, also called White Saxual, is more like a large bush than a tree with many thin trunks which regenerate well when covered by debris from seasonal flooding or by drifting sand. This makes White Saxual a good dune stabilizer. The plant can reach a height of seven meters in good conditions but usually is much smaller, gnarled, and spreading. White Saxual is not as cold tolerant as Black Saxual but is very tolerant of heat and salt, bearing temperatures up to 50° C and surviving in saline areas with only 50 mm of rainfall. It is a pioneer plant and grows in areas disturbed by sandstorms and floods. In Israel's hyper-arid Arava valley, south of the Dead Sea, saxual is known for its ability to "plant itself" on berms and in wadis. If a row of trees which do not need to be irrigated is desired, the sand is piled up in a long berm and the saxual trees obligingly appear after only a few months. White Saxual grows in wadis with sandy or gravelly bottoms and on salt flats where flood waters accumulate. Its sparse foliage is palatable for sheep, goats, and camels. The lumber is unusable but burns with the thermal efficiency of coal and makes excellent charcoal.

The Prosopis Genus of Trees

The **Prosopis** genus is an exceedingly hardy, drought tolerant, and useful family of plants with many attributes which make the Prosopis trees useful for supplying fuel, fodder, edible pods, shade, protection from winds, stopping the encroachment of sand dunes, and fixing nitrogen in depleted and arid areas. Prosopis juliflora trees have been used successfully to make ethanol, low gluten flour and syrup. In Texas wood chips and branches of the mesquite tree are held to be the best barbecue wood around. Other members of the Prosopis family are just as useful.

Prosopis alba, which is native to arid zones of Argentina, Paraguay, and Bolivia, is extremely drought tolerant. It is a tree with an attractive rounded crown from 5 to 15 meters in height which can reach one meter in diameter. They are very useful in roadside plantings and as windbreaks. Unlike many other Prosopis species, Prosopis alba has thornless variations. The young wood is yellowish in color with a slightly darker grain and brown heartwood. Prosopis alba pods can be made into baked goods and fed as a high protein and high sugar fodder to sheep and goats. The wood is exceedingly hard and not easy to work but has been used in construction, cask making, and flooring. As firewood, the heavy Prosopis limbs burn with a hot steady flame. The tree is propagated easily by seed and may be seeded directly if inoculated with mesquite rhizobia. Young trees tolerate mild frost but this is a hot weather tree and winters best in areas where temperatures rarely fall below 15° C. Trees are found in flatlands and low sierras up to 1,000 meters in warm areas.

Prosopis chilensis, also known as the Chilean mesquite, is perhaps the most productive Prosopis tree when evaluated for yield of biomass. Although native to the Pacific coast of Peru and Chile, it has naturalized itself in California, Hawaii, Africa, and India. In parts of Kenya it is considered an invasive tree. The tree can reach 8 to 15 meters in height. It has a shallow, spreading root system, greenish brown multiple trunks, and many branches. Its flowers are greenish yellow. Its seed pod is thinner than most Prosopis pods. The tree is very drought tolerant and heat-loving but it can tolerate mild frosts and very short periods of cool weather. In warm areas, it can be planted at altitudes up to 2,900 meters but it does best at temperatures between 20° C and 40° C. The tree survives with as little as 100 mm of rainfall but 200 to 400 mm of rain is needed for good growth. The wood is hard and heavy, strong and suitable for carving. It burns hot enough to fire kilns and to be used in blacksmithing.

Prosopis cineraria, locally known as "khejri," is very thorny tree. It has a long taproot which allows it to survive in desolate areas with adequate ground water. In areas such as these, Prosopis cineraria is often used as a windbreak because crop plants may be grown relatively close to this kind of Prosopis as the tree will not steal water from the protected crops. The tree reaches from 5 to 9 meters in height and has an open crown. Prosopis cineraria trees are found in the dry central parts of India, as well as, dry areas in Iran, Afghanistan, Pakistan, and the Arabia peninsula. The purplish wood is heavy and very hard. It is useful in making handles, posts, joints, and frames. As a fuelwood, it burns steadily and hotly, and makes excellent charcoal. The tree is frost hardy to -6° C and heat tolerant to 50° C. Soaked seeds sprout readily. The tree can also be propagated by suckers. The tree grows with as low as 75 mm of rainfall, even in areas with long dry seasons.

Prosopis juliflora, also commonly known as the Mesquite, is esteemed for its shade, wind breaking, and dune stopping ability. Truly a multipurpose tree, it is also highly valued as food and fodder for animals, as a fuelwood, as a source of biofuel, and for its edible products for human consumption. It originated in Central America, spread to the American southwest, and has since spread to Africa and India. This tree can reach ten meters in height under the right conditions but more often is a large, green bush. It flowers abundantly in the spring and the flowers are good sources of nectar and pollen for bees.

The leaves of the Mesquite tree are dark green and the pods are straw colored when ripe. The wood is hard, heavy, and light in color with dark veins. It burns steadily. The wood is very durable and used for door and window frames and for making posts. Mesquite trees are very heat tolerant and can bear temperature of over 50° C. Most mesquite variations are frost tolerant to -3° C, but there are mesquite types which are damaged at temperature close to 0° C. The tree's deep roots penetrate the soil in all direction in search of moisture. For healthy mesquite growth, 150 mm of rain is enough; however, the trees will tolerate areas with 750 mm of rainfall if the areas are warm. Direct seeding of the plants is possible as long as the seeds are scarified or soaked in boiling water. Mesquites grow quite quickly and aggressively, so this is a tree which should be introduced cautiously to other arid areas.

Prosopis pallida, known as the American carob, is a Prosopis type so salt tolerant it can be irrigated with water half as salty as seawater. It is heat tolerant, yet only mildly frost tolerant. The tree can reach 20 meters in height under good conditions but usually only achieves between 10 and 15 meters in areas with rainfall over 250 mm. The leaves are grayish green when dry, somewhat olive-like when green and growing. The tree is native to the drier regions of Peru, but it has spread to Columbia and Ecuador and has been introduced to Puerto Rico, Hawaii, India, and Australia. It is propagated by seeds and should be inoculated with mesquite rhizobia. The charcoal made from the tree is very useful, but in hot arid regions it is most important as a fodder tree, dune stopper, and windbreak. The lumber is rather brittle and not resistant to beetles or termites, so it is rarely used in construction. Pallida's pods are sweeter than those of most other Prosopis. Syrup and sweets can be made from them.

Prosopis tamarugo, commonly known as the Tamarugo, is a tree which survives on the salt flats of northern Chile's cold deserts. It is tolerant of cold down to -12° C and survives occasional heat waves into the 40's C. It grows from sea level to up past 1,500 meters. The Tamarugo originated in an area with very small, irregular amounts of rain and vast expanses of salt crusted earth. Large, spreading, and deciduous, it can reach 15 meters in height with trunks which are 80 cm in diameter under good conditions. In very arid areas growth is slower, but the Tamarugo has many special adaptations to aridity and salinity. The tree is able to absorb atmospheric water and establish a subsurface reservoir of moisture in the lateral root zone by exuding the moisture into the earth and reabsorbing it as needed. Roots also exude a natural herbicide which keeps other plants from growing near the Tamarugo. The wood is heavy and difficult to work, but burns well and makes good charcoal. The pods are edible and attractive to animals as is the browse and the young shoots.

Tamarugos shows promise as reforestation trees for very saline areas in other parts of the world, but many desolate areas in Chile have already been successfully replanted with Tamarugo trees.

The Tamarix Genus of Trees

These trees are among the most salt tolerant plants on earth, some hardy enough to be irrigated with sea water. Tamarix, also known as Tamarisk, taxonomy is difficult and usually can be accomplished only by experts. The plants vary widely in growth habits. Some Tamarix are bush like. Others are large spreading trees up to 20 meters tall. In general, tamarisks have fine, almost feathery looking, foliage covered with tiny scale-like leaves. The wood is heavy, hard, and durable and very slow to catch fire. Small branches and fallen leaves are too impregnated with salt to be used as kindling.

A well known hazard of areas with Tamarix trees is the tendency for the tree to salinize the area's soil by the constant drip of salt from glands in the leaves. The trees extract all available salts from the soils, excrete them on their leaves, and then return the salt to the earth in leaf litter. Another hazard in areas with large trees is falling branches, which peel off the trunk in high winds and may be both large and heavy.

These trees are especially important in erosion control and in stopping sand encroachment. Many miles of desert highways have been planted with tamarisks to keep dunes from swallowing up the pavement. Tamarisks have been planted in shelter belts in very saline areas all over the world and are used as firebreaks, because the trees and their litter are too salty to catch fire. They are not very useful as browse or fodder trees as only camels will nibble at their salty shoots, but as apiary trees, they are a generous source of pollen and nectar and they flower copiously.

The Zisiphus (or Ziziphus) Genus of Trees

Another genus of useful trees is the Zisiphus, the thorn plums. The fruits of these trees resemble small apples. Sometimes, they are mealy when green, but when they are dried, they are sweet, chewy, and easy to preserve. The wood of the trees is hard, dense, flexible, and burns with much heat and little smoke.

Zisiphus trees are among the most drought tolerant plants in the world, but they do not reach great heights. They are spreading and open-crowned in form and inclined to multiple trunks. Their wood is hard, flexible, and reddish in color. Leaves and small twigs burn well. The two species most suited for firewood are Zisiphus mauritania and Zisiphus spinichristi.

Zisiphus mauritania is sometimes called the Chinese date. It has the best fruit of the Zisiphus family with a sweet dark flesh which can be used dried, pickled, made into spreads, juiced, or simply eaten fresh. The trees produce fruits copiously. The tree is not sensitive to soil types and is extremely heat tolerant and frost tolerant. It is the most robust species of the Zisiphus family with a thick trunk, fast growth, and very heavy wood which burns fiercely and makes good charcoal. The tree coppices well, but sustainable harvesting usually is done by pruning off the lower branches and letting young wood bear fruit.

Zisiphus spinichristi is a small spiny desert tree found in very dry areas. The tree survives where water gathers in wadis and small depressions in areas with less than 100 mm of rain. The fruit is widely variable in form, but usually is round, small, and mushy when just ripe, but of excellent flavor when dried on the tree. In the Middle East, it is a favorite of hikers, animal herders, and soldiers because it remains edible for many months. Sheep, goats, and camels enjoy fallen fruits and young shoots of the plants. The wood is naturally termite proof. It is used for handles of tools, posts, and household utensils. It burns hotly and makes good charcoal. The tree can be found from West Africa to the Red Sea. It is also found in Israel, Jordan, Turkey, Iran, and the Saudi Arabian Peninsula.

The aforementioned trees can be planted in some of the hottest and most desolate places on earth to provide shade, fruits, browse, fodder, and fuel as well as helping to restore damaged soils and ecosystems. The fuel needs of the earth's poorest people can be met while restoring and reviving the places where they live.

Chapter 7
Arboreal Pastures

One of the more absurd notions about widespread hunger in the world is the idea that everything would be all right if we all stopped eating meat and drinking milk. Production of meat and milk, the reasoning goes, uses up scarce grain resources which could be fed instead to the poor. The methane from the animal waste products is helping to heat up the planet and cows are the "Hummers" of the agricultural world". Supposedly vegetarians leave a smaller carbon footprint on the earth. Therefore, continuing the thought, we all ought to be vegetarians. This, of course, is a totally unreasonable notion for several reasons.

First of all, only wealthy, overdeveloped, nations feed their livestock grain. Everywhere else the livestock eat grass, hay, straw, food scraps, agricultural waste, wild browse, and fodder.

Secondly, it is debatable whether the methane from the digestive systems of most animals is fueling climate change since grass fed cows burp out much less methane than do cows fed on corn and soybeans. (Most of the methane actually does come out of the front end of the cow.)

Lastly, a vegetarian diet actually requires more prime cropland and sweet water than a diet which contains meat and milk.

If one leaves out the extravagant folly of feedlot steers, battery birds, bovine growth hormone saturated cows, and pigs raised in crush pens and if one returns to traditional forms of food production, it is possible to see the farm animals not only improve the diet of the farmer and the farm family, but also utilize food resources which the human inhabitants of rural areas cannot. These animals are also critically important to the nutrient cycling which keeps farmland productive. Indeed, the scantier and more difficult the area, the more likely it is the people who live there will have and keep domestic animals for milk and meat, hair, hides, and manure.

Examples can be found on every continent: the Lapps herd and follow the reindeer, the Bedouin and the Tuaregs keep goats, sheep, and camels, the Navaho herd sheep and the Ma'asai make their living from African cattle. In the Andes, guinea pigs are important sources of protein while llamas, vicunas, and alpacas supply fibers, hides, and meat, and also assist in transportation. In Tibet and Nepal the yak is the most important domestic animal. At the top of the world, the musk ox is a source of fiber and food.

If these animals were such a burden to the local food supply, why would people from such challenging environments keep them? The answer is simple: Only the unsustainable practices of modern animal husbandry pit animals against people in a competition for food and grain. Only the unsustainable practices of the feedlot turn animal manure, a precious source of renewed fertility, into toxic waste. Only the practices of modern animal husbandry turn healthy sources of nutrition, such as meat and milk, into clogged arteries and damaged hearts. Sensible, traditional methods of animal husbandry feed animals on substances unsuitable for human consumption and use the animal products and the animals themselves for a great variety of purposes.

The people of marginal areas and challenging environments cannot conceive of life without their animals. One of the most successful poverty mitigation programs of all time, Heifer International (http://

www.heifer.org/), in recognition of the great value of animals to farms and to agrarian families, has as its main purpose the distribution of calves, sheep, goats, chickens, ducks, and geese. The people of Heifer International see the gift of these animals as a way to help people out of abject poverty. This amazing program gives encouragement to very poor rural families by providing training in animal care and pairs of breeding animals. The only condition is some of the offspring from these animals must then be gifted to other local families.

How irritating then, is the chorus from the overfed nations of the west, preaching vegetarianism as a way to stretch the world's food supply when it is no such thing.

The first and easiest way to stop the competition between domestic animals and the world's poor is to stop feeding the animals grain. Grain- fed animals are generally unhealthy animals because they were made too fat too soon, and crowded into nightmarish feeding lots, then dosed with antibiotics and chemicals just to keep them alive until slaughter. Domestic animals, even chickens and pigs, are much healthier when put to pasture and can spend most of their lives and get most of their food from a good mixed pasturage. Traditionally an animal was only confined and fed a richer diet for a short period before slaughter. Therefore, competition between the world's poor and domestic animals for staple foods like wheat, rice, soybeans and corn is a very recent development.

New ecologically minded animal feeding operations have returned to the practice of grass feeding domestic animals with truly encouraging results. For example, the Polyface farm, described by Michael Pollen in the *Omnivore's Dilemma*, is a successful rotational grazing operation. Recognizing the fact the simplest way to capture the sun's energy is to grow grass, and then use the grass as a feedstock for a variety of endeavors, the owner of the farm, Joe Salatin, rotates cows, pigs, turkeys, rabbits, and chickens through a series of electrically fenced paddocks and pastures. He also raises tomatoes, sweet corn and berries. When asked what kind of farmer he considers himself to be he says, “I'm a grass farmer” because grass is the basis of an intricate food chain.

The animals mentioned consume the grass directly or indirectly but they also aid in the regeneration of the pastures. For example, hay will be cut from a field already grazed by beef cattle and then picked over by several hundred chickens brought to the pasture in a movable coop. The chickens perform several important ecological functions including eating flies and fly larva from the cowpats, spreading the manure, and picking up other insects. They also eat grass and deposit their own high nitrogen manure on the pasture. All the while, the chickens are gaining in weight and producing eggs. After the first cut of hay and a few weeks respite from grazing animals, the pasture will be used again by beef cattle. Hay will be cut again at the end of the summer. The hay will be stored over the winter and fed to a variety of animals. By the end of the year, the grasses will have been transformed into beef, pork, broiler chickens, rabbits, eggs, and turkeys, all with no loss of fertility to the pasture.

This is not a surprising outcome. Natural grasslands are kept in balance by the presence of herd animals and cannot regenerate without the animal's feces and urine to return fertility to the soil and the action of the animal's hooves to tread in the seeds of wild plants and grasses. Land fenced off in African wildlife reserves in a 1995 experiment did not recover its vegetation. Grass did not regrow and trees and bushes did not begin to sprout until the herd animals were allowed to pass through the area. Grasslands need grazing animals as much as the animals need the grass.

Biomimicry of grassland ecologies in the new grass feeding farms, like Polyface Farm, allows the land to benefit from the nutrient recycling of the grass fed creatures and their natural proclivities such as grazing, rooting, and insect eating. So pigs are encouraged to root up paddocks and chickens are used for maggot and fly control in areas where cattle have been grazed.

Joe Salatin's book about pastured poultry, *Pastured Poultry Profits*, published in 2008, explains the system in detail and outlines the great benefits conferred by farming animals in pastures on some of the 450 acres of his farm. Polyface Farm produces all of its many products with a grass diet and a minimal investment in off farm inputs. It can be a model for many temperate zone farms as it was developed on reclaimed land, it is profitable, it is sustainable, and it is based on a system which requires careful management and ingenuity instead of expensive fertilizer inputs.

If domestic animals are to be returned to pasture it must be done in a way which increases the amount of food for animals without putting more strain on agricultural systems. In marginal areas, often used for grazing, it is possible to put problems together to make solutions. This is the case with arboreal pastures; trees planted with the dual purposes of controlling erosion and regenerating depleted land *and* providing food, shade, and browse for domestic animals.

Tree forages can contribute biomass of high quality and high digestibility. They can supplement a nutrient deficient diet and enhance the microbial growth and action in the rumen, supply supplemental protein, contribute vitamins and minerals not present in the basal feed resources and so reduce or eliminate the requirement for expensive purchased concentrates. All this, of course, allows for a decrease in the cost of feeding the animals. Arboreal pastures are particularly effective in arid, saline, marginal, "worked out," deforested, and depleted areas where people who have little but their animals must make their living from what remains of the indigenous vegetation.

In the deserts, mountains, and marginal areas of the world, different groups of people often compete for the same resources, especially relating to grazing rights. Such a conflict, known as a "shepherd's war" has often led to serious violence, blood feuds, and murder. The story of Cain and Abel may be the earliest recorded account of such a dispute, but in the recent past of the US there were violent conflicts between cattlemen and homesteaders, and cattlemen and sheep herders. In many areas where resources are scarce, feuds and conflicts over water and vegetation still occasionally break out into violence. In the end, these rivalries center on use of water resources and access to areas of natural vegetation where animals can graze.

The conflict over water can usually be resolved by mutual agreement as to when and where the flocks may drink from natural water sources. However, vegetation, lush in the spring, becomes sparse and thin in the summer, absent in the winter, and may be destroyed totally by overgrazing. To this end, the use of runoff, flood, and non-sewage waste water to promote the growth of indigenous wild trees for grazing and to enhance the quantity of native perennial plants with value as fodder can increase the resource base and thereby engender cooperation and ease tensions between communities. This type of agroforestry is dependent on efficient use of previously untapped water resources, regional cooperation, and the planting of hardy native perennials chosen for their usefulness as reclaimatives, browse, and fodder. Programs to plant trees to serve as arboreal pasture plants have been successful in Mali, Niger, Kenya, Ethiopia, and Israel.

In these successful programs, tree species were chosen for their hardiness, the type and amount of food they supply for the flocks, their ability to regenerate, and often for their soil building qualities such as nitrogen fixation. Early experimentation allows researchers to discover the number and type of trees per head which are needed to support domestic animals. The animal herders are taught to move the flocks often to allow the trees to regenerate. Plantings are mixed to avoid problems with the "ration." Mixed plantings allow animals to choose the food and fodder most appealing to them at that time and season of the year. This avoids the problems of overfeeding of specific plant materials which can result in problems with fertility (from overfeeding carob pods and leaves) to sore mouths (Prosopis pods and fodder) to digestive problems (overfeeding on oak mast). In mixed plantings these problems do not occur as the animals will instinctively vary their diets if the opportunity presents itself.

In some very dry areas in Niger, local people are being taught to think of Acacia trees not as raw material for charcoal production but for their value as pasturage for their traditional flocks of sheep goats and camels and also as feedstock for a new domestic animal, the silkworm. Further south on the African continent are programs which encourage the "mini pasturage" of snails and mopane worms on native trees. There is even a chickenfeed tree project, using the fruit and flowers of a Croton megalocarpus species.

One interesting project, an example of dryland agroforestry, in Israel near the Bedouin village of Hura and sponsored by the Nevada based International Arid Lands Consortium (IALC) has planted hundreds of trees on a farm owned by the Abu Rabia family. Receiving only 200mm of rain, the areas nearby tend to be severely degraded due to the overstocking of poultry and sheep. Largely unused areas of hills and valleys, suitable for terracing and water harvesting, and following natural watersheds were dammed for water collection. After this, a variety of trees were planted including Prosopis species, olives, acacias, carobs, arganias, cassias, pistachios, and various drought tolerant fruit trees. (The project is fully described in the journal Management of Environmental Quality Volume 19, no. 3, 2008. See Appendix II of this book.) A tenfold increase in fodder is expected as well as supplies of fruit, firewood, nuts, poles, and mulching material. This greatly increases the farm's productivity and brings in much needed income from the increased amount of milk and cheese produced by the herd animals.

This type of project could be of great value in the dry and marginal areas of the world. Degraded drylands under well planned rehabilitative planting strategies have the potential to produce immense amounts of food and biomass and store vast amounts of carbon. Such strategies would help the marginalized nomadic peoples of arid areas to regain their independence and safeguard their traditional ways of life while mitigating their poverty and increasing the carrying capacity of these depleted areas.

In high altitude grazing areas tree forages already form a high percentage of food for ruminants. The use of tree forage is widespread in tropical areas as well. In these places there are well established systems of animal farming which integrate seasonally available grass, agricultural waste, and tree fodder to support the flocks and herds year round.

In order to determine which plants and shrubs are suitable for use as components of the animals' diets, the local vegetation of a given area must be investigated and evaluated. Some of the considerations are: the capacity of the plant to regenerate when grazed, the feeding behavior of the animals, voluntary intake of the fodder under different conditions, the nutritive value of the foliage, and the year round availability of forage. Before fodder trees are introduced, it is wise to evaluate the imported candidate trees for weedy characteristics, determine the pH and soil requirements of these new trees, and determine if the growth pattern of the trees fits well with the seasons and the cycles of other trees and crops. The use of introduced species is sometimes necessary but usually there is sufficient variety, even in relatively sparse ecosystems, to find and develop suitable grazing trees.

Trees for a Variety of Climates and Zones

Below is a list of the most popular grazing trees in the hills of Nepal, an area both cold and seasonally quite dry.

Albizia julibrissin var. mollis
Artocarpus lakoocha
Bauhinia purpurea
Buddleja asiatica
Castanopsis hystrix
Celtis australis
Dalbergia sissoo
Dendrocalamus strictus

Erythrina arborescens
Eurya acuminata
Ficus nemoralis
Grewia oppositifolia
Grewia tiliaefolia
Litsea polyantha
Machilus odoratissima
Michelia champaca
Prunus cerasoides
Quercus lamellosa
Quercus semecarpifolia
Salix spp.
Saurauia napaulensis
Schima wallichii
Ulmus wallichiana

Some of these trees are familiar trees of the temperate zones, put here to another type of use. If *Quercus spp.* (the oak trees), *Salix spp.* (the willow trees), *Prunus spp.* (the wild almond trees), and *Ulmus spp.* (the elm trees) are suitable and hardy for grazing in Nepal, then perhaps they should be candidate species for tree-based fodder in reclaimed areas in the temperate zones.

Other Temperate Zone Fodder Trees

There are many suitable fodder species in temperate zones, most of them abundant enough to make a difference in the available ration if managed properly. Tree leaves are generally far richer in dry matter, nitrogenous matter, and particularly trace minerals than grass or alfalfa. However, their high lignin content, their tannin content, and sometimes their astringent properties make leaves less digestible than most cultivated fodder plants.

It is also a fact that tree fodder (leaves, pods, and small twigs), if fed exclusively, can be problematic for domestic animals. As examples of the range of issues, exclusive feeding on carob leaves and pods severely inhibits animal fertility. Feeding exclusively on *Prosopis spp.,* despite the high quality of the protein rich pods and high quality browse, may lead to sore mouths and digestive problems. The leaves and small branches of *Salix spp.* (willows), *Alnus spp.* (alders), *Castanea spp.* (chestnut trees), *Aesculus spp.* (buckeye or horse chestnut trees), and *Olea europaea* (olive trees) are difficult to digest and are unappetizing to cattle as a rule, but more suitable for sheep and goats. Coniferous needles are absolutely indigestible and unappetizing. *Populus nigra* (Black poplar) leaves are toxic. *Juglans spp.* (walnut trees), *Laburnum spp.* (Golden Chain laburnum), *Cytisus scoparius* (scotch broom), *Taxus spp.* (yew), and *Buxus spp.* (boxwood) leaves are very poisonous for animals (and humans, for that matter). Feeding olive leaves in quantity gives a very bad taste to the milk of ruminants and can cause liver problems in sheep.

Because of issues like these, tree fodder from a variety of trees is more beneficial than depending on a single local species. The exceptional richness of fodder leaves from the better local species compensates partially or even completely for their lack of easy digestibility. Varying the type of tree fodder eliminates the other problems.

The leaves from fodder trees can contain as much as 0.35 Feed Units per Kg (One FU per kilogram equals.78 pounds of digestible material) of green matter and 18 to 20% of nitrogenous matter which combine to give a nutritional value of complete nitrogenous proteins almost twice as high as clover or alfalfa. However, alfalfa proteins and those of clover are absorbed more easily by cattle than those of most tree fodders. Careful planning may be required to balance nutrition and digestibility in the ration and to allow the instincts of grazing animals to draw them to the more suitable feed.

Although the nutritional value and digestibility of tree leaves varies with different species, some species give fodder of high nutritional value which is particularly appreciated by cattle, for example *Ulmus spp.* (elm trees), *Morus spp.* (mulberry trees) and *Sambucus spp.* (elder trees) which furnish excellent leaf material. There are other trees, such as *Betula spp.* (birch trees) and *Quercus spp.* (oak trees), which are much relished in the spring but are not as pleasing to the animals in the summer and fall, probably due to the rise of tannin content. The leaves of *Fagus spp.* (beech trees), *Populus spp.* [poplar trees, except *Populus nigra* (Black Poplar) which is toxic], *Corylus spp.* (Hazelnut trees), *Sorbus aucuparia* (European Mountain Ash), *Betula pendula* (Silver birch), *Tilia spp.* (linden trees), *Ficus spp.* (Fig trees), and *Chamaecytisus proliferus* (Tree Lucerne or Tagasaste) have sufficient nutritive value to serve as fodder for cattle, sheep, and goats as well being attractive to the animals through most of the year.

Oak and evergreen oak leaves are appropriate for sheep and goats but allowed only in very small quantities for cattle. Feeding these leaves reduces the milk production in cows. They are toxic because of their high tannin content. Oak leaves in quantity should only be fed to dry sheep and never to milk producing animals. However, oak acorns are good for livestock because of their richness in feed units at the end of autumn and during winter. Oak mast (leaves and acorns) is the traditional forage food for fattening pigs. Also, acorns may be crushed and roasted for feeding some fowl.

Ash is another attractive leaf forage plant. The ash tree's dried leaves were used as reserve winter fodder in the mountains of Laucaune (Montague Noire, southern foothills of the Massif Central) where some ruminants seem to prefer it to hay.

European Marine Gorse (Ullex galli)is an interesting fodder plant for temperate zones and has great unrecognized potential. This leguminous plant produces a green fodder of high nutritional value which livestock prefer to hay. It gives peak fodder production from the end of November to the end of February which can be harvested just in time to replace the biomass from exhausted autumn pastures. Traditional livestock feeding in Brittany and Wales was based on gorse. Sheep, goats, cattle, and horses were fed on ground or chopped gorse sprouts during the entire winter period. The entire three to four meter high plant was cut to the ground every two years and allowed to regrow. The top of the plant was used for fodder. The rest of the plant was used as animal bedding or firewood. Records show the plant yields an average of 50 to 100 tons of Dry Matter material per hectare per year. Because its maximum yield is in the middle of winter, just when the pastures have nothing to give, gorse culture could constitute an important link in the fodder chain. Gorse is a leguminous plant which is valuable for developing sandy and acidic soils, reclaiming depleted and uncultivated land, and can be wildcrafted from very rough areas unsuited for cultivation where it produces an average of 15 tons of Dry Matter per hectare per year. It is a valuable feed since 12 kgs. of gorse equals 8 kgs. of hay and 4 to 5 kgs. of oats.

Other small tree/shrubby species for rough areas include European buckthorn (Rhamnus cathartica) which is suitable for cool montane rocky areas, and *Hippophae rhamnoides* (common seabuckthorn) which is suitable for very dry areas, and also *Erica spp.* (heather) which are suitable for cool, wet, and acidic soils. European buckthorn has become invasive in North America so it is a plant that should be handled with care.

An interesting paper by Sabrine Karg of the national Museum of Denmark describes the uses of heather in Bronze Age societies with heather plants being cut and stored like hay for winter feeding of animals and heather sods used for building. This pattern of use established itself in northern England, Scotland, and Ireland and also other areas with acid soils. Often, a periodic burning off of the heathland was done to eliminate less desirable plants and to allow the heather to renew itself.

The nuts of *Aesculus spp.* (buckeye or horse chestnut trees), often called conkers, are traditionally used as reserve food and fodder and are digestible if crushed or ground. Many domestic animals were kept alive through the very lean years of WWII in Europe by biomass gleaned from the horse chestnut.

The genera *Gleditsia* or *Robinia* should not be neglected as some of the best tree forages come from them. *Gleditsia triacanthos* (Honey locust) is perhaps the best of them with nutritious leaves and pods full of sugar and protein which are irresistible to many domestic animals.

Leaf forage can be used as fodder condiments because it is rich in trace minerals which are brought to the surface by the deep roots after dissolution of the bedrock. That is why tree leaves are much richer in trace elements than grass. So a regular distribution of a small quantity of leaves can be very good for the livestock's health. Leaf fodder can be very useful for stretching out hay reserves during winter. Thus elm leaves were sometimes collected into dryleaf faggots for sheep and goats in the Massif Central and in Italy. Finally, forage trees can serve as reserve fodder with a view to filling a gap in summer, as well as emergency pastures in case of heavy droughts. It does not make economic sense to cut hay in spring only to feed it to the livestock in summer and so be deprived of good quality feed during the seasons the animals need it the most.

NFT-Nitrogen Fixing Trees

Nitrogen-fixing trees, useful for land reclamation and animal feeding, are suitable for many challenging environments. Many of them are multipurpose species, producing timber, firewood, syrup, oil, and wood for making tool handles or carving.

Here are some examples for consideration:

Acacia aneura - a fodder tree for sandy hot deserts

Acacia auriculiformis - a multipurpose tropical wattle, fodder, firewood, and timber

Acacia gerradi - shade, timber and livestock food for hyper-arid zones over 600 m

Acacia koa - Hawaii's most valued native tree

Acacia leucophloea - shade and fodder for livestock in arid environments

Acacia mangium - an important multipurpose tree for the tropic lowlands

Acacia nilotica - pioneer plant for dry lands; salt tolerant

Acacia radiana - shade and livestock food for hyper-arid, hot zones

Acacia saligna - for dryland fodder and soil stabilization; can be invasive

Acacia senegal - gum tree with promise for agroforestry and grazing
Acacia seyal - multipurpose tree of the Sahara desert, also produces gum
Acacia tortilis - fodder tree for desert sands at low altitudes
Adenanthera pavonina - an underutilized tree of the humid tropics
Albizia odoratissima - shade tree with soft palatable fodder and pods
Albizia procera - for reforestation and agroforestry
Albizia saman - pasture improvement, shade, timber, and more; for warm zones
Alnus acuminate - valuable timber tree for tropical highlands
Alnus nepalensis - a multipurpose tree for the tropical highlands
Andira inermis - a beautiful ornamental tree with value as food and fodder
Casuarina equisetifolia - use regularly for windbreaks and land reclamation
Casuarina glauca - a hardy tree with many attributes
Casuarina junghuhniana - a highly adaptable, edible tropical Casuarina
Chamaecytisus palmensis - hardy, productive fodder shrub
Dalbergia latifolia - the high-valued Indian rosewood multipurpose species
Dalbergia melanoxylon - valuable wood from a neglected tree

Enterolobium cyclocarpum - the ear pod tree for pasture, silage, fodder, and wood
Erythrina edulis - multipurpose tree for the tropical highlands
Erythrina poeppigiana - shade tree gains new perspectives
Erythrina sandwicensis - unique Hawaiian NFT
Erythrina variegata - multipurpose tree
Faidherbia albida - particularly suited dryzone agroforestry
Gleditsia triacanthos - Honey locust: widely adapted temperate zone fodder tree
Guazuma ulmifolia - widely adapted tree for fodder
Hippophaë rhamnoides - an NFT valued for centuries; produces edible fruit; pioneer plant in cool arid areas
Inga edulis - a tree for acidic soils in the humid tropics
Juliflorae acacias - new food source for the Sahel and other degraded grazing lands
Leucaena diversifolia - fast growing highland NFT species
Leucaena leucophyla - an important multipurpose tree with a soft palatable fodder
Olneya tesota - a potential food and fodder crop for hot arid zones
Ougeinia dalbergioides - a multipurpose tree for sub-tropical and tropical mountain regions
Pentaclethra macrophylla - a multipurpose tree from Africa with potential for agroforestry in the tropics

Pongamia pinnata - a nitrogen fixing tree with valuable oilseeds
Pithecellobium dulce - sweet and thorny
Prosopis alba - subtropical semi-arid fuel, food, and fodder tree
Prosopis chilensis - subtropical semi-arid fuel, food, and fodder tree
Prosopis cineraria - a multipurpose tree for arid areas, food, fuel, fodder, and dune stabilization
Prosopis glandulosa - Honey mesquite: a multi-purpose tree for arid lands
Prosopis juliflora - mesquite, suitable for arid areas with poor soil
Prosopis nigra - multipurpose species for hyper-arid zones at altitude
Prosopis tamarugo - cold saline deserts
Pterocarpus indicus – the most majestic nitrogen fixing tree
Robinia pseudoacacia - temperate legume tree with worldwide potential and exceptionally hard wood
Sesbania sesban - widely distributed multipurpose NFT, very adaptable
Sesbania grandiflora - NFT for beauty, shade, food, fodder and soil improvement
Ziziphus spina-christi - fodder and edible fruit for arid, saline zones

There are many tree species with which to work. Appropriate species can be grown in just about any climate. Tree forage can sustain and nourish much of the world's population of meat and milk animals. This can be accomplished without taking as much as a single mouthful of grain from a hungry child.

Chapter 8
Trees for Edible Oil

Edible oil is considered a necessary staple food. In Biblical times if one had grain for bread, wine to make the heart merry, and oil to make the face shine—one was considered a fortunate and well fed person. The nutritional value and the food energy supplied by oils and fats are often misinterpreted or undervalued in the human diet. Also undervalued is just how necessary edible oils are in the preparation of other kinds of food.

Modern ideas about fats and oils are often faddish and trendy. Animal fats are made the scapegoats for a host of diet related diseases of the modern era when they are no such thing. Worse, potentially dangerous "transfats" are escaping scrutiny and are being sold as miracle substances.

Inexplicably, while some traditional oils, such as olive oil, are enjoying resurgent popularity, at the same time, other traditional oils are being replaced by soya, corn oil, and rape seed oil. All three of these replacement oils are problematic for the ecosystem and the human body when produced and consumed in quantity. More disturbing than this are the organized efforts to destroy local edible oil industries, which use local and ecologically appropriate oil seeds, and replace them with inferior commercially made oils. In too many places the efforts are successful. Now areas of the world which were once self-sufficient in edible oil have become dependent on the global market for the all-too-common corn, rape seed, and soya oils.

The most disturbing trend of all is the invention of substances such as "Olestra"which acts like a lipid but cannot be utilized by the human body. As it passes through the digestive system, it picks up minerals and vitamins, resulting in the first recorded instances of "negative nutrition." In other words, someone invented a "food" which impoverishes rather than nourishes the body.

Rather than enter into the controversies of the "oil wars" between scientists, nutritionists, and other interested parties, this chapter will examine the idea of increasing the edible oil supply by providing more oil from arboreal sources, as well as increasing the diversity of available edible oils available on the market.

Old Favorites

These are tried and true edible oils with many advantages over modern commercial oils and many special qualities.

Walnut Oil

Walnut oil is pressed from the nuts of a large deciduous tree which grows in the temperate zones of the world and at high altitudes. The large green hulls split into four segments when the nut is ripe. The hulls are sometimes used to dye cloth. Walnut oil was one of the earliest oils used in painting but in the last century, it has been used mostly in food and cosmetics preparations and occasionally in carpentry to stain wood. The English or Persian walnuts are the most popular sources of walnut oil, although the wild Black walnut's resinous oil has also been used for cooking. Roasted walnut oil is prized for its rich topaz color and its delicate flavor but unroasted oil is considered to have desirable anti-aging properties. Walnut oils often have different profiles depending on where the trees are grown but generally the oil content is 50% to 70% with the major fatty acids being Oleic 14-21%, Linoleic 54-65%, Palmitic 6-8%, and Stearic 1-3%.

Almond Oil

The Almond tree is deciduous with green and silver leaves 3 to 6 meters in height and bearing a small fruit which looks like a green peach. Almonds are found in temperate and Mediterranean zones all over the world. The outer layer of the fruit is thin, tough, and inedible. The fruit splits open when it is ripe but the pit contains a large kernel. Almonds can be divided into two types, sweet almonds and bitter almonds. Almonds are also divided into hard shelled and soft shelled types. Sweet almonds are widely used as food, either as whole nuts or ingredients in cooked or processed foods. They are generally too valuable to be pressed for oil but their cold pressed oil is sometimes sold as a gourmet item for baking or cooking and its intense coconut like flavor is highly esteemed. This product is sold as Food Grade Sweet Almond Oil. Sometimes almonds are roasted before pressing into edible oil, producing oil with a very different taste. The nutritional content of Almond oil is 78% fats, 12% protein, and 10% carbohydrates.

Bitter almonds contain Amygdalin and enzymes which cause its hydrolysis to glucose, benzaldehyde, and hydrocyanic acid. It is these almonds which are pressed to produce both Sweet Almond oil, a fixed oil used in cosmetics, and the steam extracted volatile oil of Bitter Almonds.

Sweet Almond oil is used in many cosmetic products. The oil cake after extraction contains between 37% and 47% protein and at least 10% oil. This oil cake is used in animal feed. The major fatty acids in almond oil are as follows. Palmitic acid 7.5%, Stearic acid 1.8%, Oleic acid 66.4%, Linoleic acid 23.5%.

Olive Oil

Olive trees, originating around the Mediterranean, are now grown in warm and dry climates all over the world. Recently, the trees have been planted in Africa and South America as well as Australia with an estimated total of 800 million cultivated olive trees worldwide. In Greece, 60% of the arable land is devoted to olive orchards. The best olive oil in the world still comes from the countries around the Mediterranean Sea. Pressed from the ground-up fruit pulp and seed pit, olive oil is one of the most stable oils. The pure oil rarely becomes rancid and can be kept for many years in cool dark places. It is considered the best of the monounsaturated fats and is a prime component of the "Mediterranean Diet" concept. It is one of the few oils which can be consumed as it is freshly pressed. Indeed, the first press of the oil, low in acid and high in Vitamin E is highly esteemed and sold as virgin and extra virgin olive oil.

Olive oil is "healing" oil, both inside and outside of the human body. Inside, it lowers bad cholesterol in the bloodstream, heals gastric problems, protects cardiac health, and reduces cancer risks. Outside, Olive oil soothes chapped or sunburned skin and is non-drying in soaps, lotions, shampoos, and creams. Because Olive oil tastes strongly of the fruit from which it is pressed, it is sometimes considered too pungent for frying. However, it is the most popular oil for salads and it is often combined with other ingredients to make sauces, salad dressings, and gravies. The major fatty acids in Olive oil are: Oleic and Palmitic acids.

Edible Olive Oils

Virgin, Extra Virgin, Ordinary Virgin: what are they talking about?

Virgin Olive Oil is oil obtained only from the olive, using solely mechanical means which do not alter the oil. It has not undergone any other treatment besides washing of the fruit and the decanting, centrifuging, and filtering of the oil. No heat or solvents or chemical means of oil extraction have been used. Virgin olive oil has a free acidity expressed as Oleic acid of not more than 2 grams per 100 grams.

Extra Virgin Olive Oil is a sub-class of Virgin Olive oil which has a free acidity expressed as oleic acid of 0.8 grams per 100 grams. This is the best grade of olive oil with the finest taste. It is used for soups, stews, and "dipping," which is, eating bread with the Olive oil. Extra virgin olive oil accounts for the best 10% of the virgin olive oil produced.

Ordinary Virgin Olive Oil is pure but with more acidity, about 3.3 grams per 100 grams and a strong taste. Ordinary virgin oil is excellent for cooking and frying but considered rather harsh for salads and dipping. It is used in soap making for the better grades of olive oil soap.

Refined Olive Oil has an acidity of about 0.3 grams per 100 and this has been achieved by filtering by the use of charcoal and other chemical means. No solvents have been used for extraction.

Olive oil (with no qualifier) is usually a mixture of cheap refined oil and virgin olive oil, blended to give the inferior oil some flavor. The acidity averages out to 1 gram per 100 grams.

Inedible Olive Oils

Olive Pomace oil is extracted from the olive waste and not fit for human consumption being used mostly in soap making and industrial processes.

Lapante Virgin Oilve oil is Virgin Oil not fit for consumption because of high level of acidity (3.3 grams per 100), impurities, etc. It will be refined or used industrially.

Pine Nut Oil

Pine Nut oil is made from the seeds of several species of pine including Stone pines from Europe, Korean pines and Chilgoza pines from Asia, and Pinyon pines from North America. Pine Nut oil is an excellent preservative. It is also a valuable medicinal oil which suppresses the appetite, reduces LDLs, and helps heal ulcers and gastritis. Pine Nut oil has a low smoke point, however, it is generally not used for cooking, but it is added to cooked food to finish it and to keep it fresh. In Siberia, a handful of pine nuts or a tablespoon of Pine Nut oil was taken with a slice of bread when food was scarce to induce a feeling of satiation. Pine nuts are used in many foods world wide, being added to rice, fish, meat, or vegetable dishes. Pine nuts are also eaten roasted and/or salted. They are often used in baking and desert making, being added to chocolate or being ground-up and used with pistachio nuts in Baklava. The major fatty acids of pine nut oil are: Linoleic acid 49%, Oleic 24%, Pinolenic acid 17%, Palmitic acid 6-7%, Stearic acid 2.5%

Hazelnut Oil

The hazelnut tree, sometimes called the filbert tree (Corylus avellana) is cultivated in Italy, Spain, Turkey, Portugal, Greece, and France though most of the Hazelnut oil on the market is from the first three countries. A very healthy oil, it is often recommended to people who suffer from problems with cholesterol. It has a distinct, pleasant and delicate flavor. It is made by cold pressing shelled hazelnuts. While the oil is considered dietetic and medicinal, it is also used in baking and candy making. The major fatty acids in Hazelnut oil are: Myristic 14%, Palmitic 16%, Palmitoleic 16%, Margaroleic 17%, Stearic 18%, Oleic 18%, Linoleic 18.2%.

Apricot Oil

Apricot oil is similar to Almond oil. Since it is much cheaper than Almond oil, it is sometimes used to cut Almond oil and sold to be used in the making of cosmetics, soaps, and perfume. It can be used on the scalp to improve the health of the hair and skin. and is considered one of the more soothing oils for chapped and broken skin. It is cold pressed from apricot pits, which contain 40% to 50% oil. Its major fatty acids are Oleic and Linoleic. The Apricot oil, while considered edible by some, is usually not used in food because of variable levels of toxicity from the presence of Amygdalin which imparts a residual bitterness of taste to the oil.

Avocado Oil

Avocado trees grow well in subtropical, tropical, and some temperate locales. They are well known for their large green and brown skinned fruit, a fruit with buttery flesh and a large seed. Avocado oil is pressed from fruit which has been left to ripen on the tree instead of harvested green for human consumption. The trees may reach 20 meters in height and produce fruit from five years of age to 30 years of age. The oil produced from the fully ripened fruit has been traditionally produced by mashing the pulp and seeds thoroughly, rendering the resulting slurry, and skimming off the oil. Then the oil is whirled in a centrifuge until the oil separates from any remaining impurities. Avocado oil is sometimes refined further by filtration. The resulting oil is a high grade edible one, but avocado oil is used increasingly in cosmetics and as sunscreen, where it is particularly effective. Avocado oil is not used for cooking as it becomes bitter when heated but it is used as a superior salad and "finishing oil." Avocado oil composition varies according to race and variety of the Avocado tree, but the major fatty acids involved are Palmitic 7.2% to 26%, Oleic 64% to 80%, and Linoleic 6.3% to 11%.

Lost Crops

These oils, once highly valued, were replaced by cheaper industrially produced oils or became rare as stands of the trees used to produce the oil diminished. Several of them are making a comeback.

Cashew Oil

Cashew oil is pressed from grade B cashew nuts. Both oil and nuts are excellent food containing many trace elements. These popular edible nuts, usually sold pre-roasted and salted, are actually the "tail" end of a fruit called the cashew apple. These soft red and purple fruits are pressed for juice which is sold fresh and sweetened or fermented into liquor. The nuts themselves contain a large amount of anacardic acid, and they must be dried and roasted to be safe for consumption. The large tropical Cashew tree (*Anacardium occidentale*) which produces both nuts and fruit, is a relative of the Mango (*Mangifera indica*) and the Marula (*Sclerocarya birrea*). All three trees are in the same tree family: *Anacardiaceae*. The major fatty acids in Cashew oil are: Oleic 73.3%, Linoleic 7.67%, Palmitic 0.89%, Stearic 11.24%, Lignoceric 0.15%.

Macadamia Oil

Macadamia oil was once a rare locally hand pressed oil in Australia. Now, commercial Macadamia oil is pressed from two species of the hard-shelled nuts (*Macadamia tetraphylla* and *Macadamia integrifolia*) which are mostly sold as a high quality and high priced edible nuts. Macadamia nut oil is one of the finest oils for cooking as it is stable and does not go rancid easily, while imparting the taste of whole macadamia nuts to the food which is cooked in it.

The Macadamia tree is native to the coastal rainforests of Queensland but is now cultivated in many countries in Africa and Asia. The trees are medium sized with dark bark and slightly prickly dark green leaves. The trees cannot tolerate frost and do best where there are alternating wet and dry seasons. Veteran plantations of Macadamias exist in Australia and Hawaii, which are where most crop research is taking place. Trees begin to bear after 6 to 7 years. Most commercial trees are of the second species *M. intergrifolia*. Nuts develop slowly and fall 215 to 230 days after flowering. This occurs between August and February in the northern hemisphere and between March and September in the southern hemisphere.

Macadamia nuts are gathered on the raked ground by hand, or are allowed to fall into nets or spread canvas, or they are recovered after they fall by tractor-powered recovery systems, including vacuum collectors. The husks must be removed immediately and the nuts dried to prevent spoilage. Nuts should be dried where there are temperatures under 40C and good air circulation. The shells of the nuts are used as mulch and fuel, while the oilcake is fed to non-ruminant animals after pressing. Hull powder is useful as filler in the plastics industry. The cracked nuts are graded and the best are sold as edible nuts. The other nuts are ground and oil is extracted by pressing and expelling. The major fatty acids of macadamia oil are: Myristic acid 0.7%, Palmitic 9.1%, Palmitoleic 21.9%, Stearic 2.2%, Oleic 59.9%, Linoleic 1.9%, Arachidonic 1.8%, Eicosenoic 2.0% making it one of the richer and more complex edible oils.

Argania Oil

Argania oil is one of the harder oils to obtain. It is a rich, golden oil from a little known tree from Morocco. The Argan tree (*Argania spinosa*) is also called the Ironwood tree. These trees grow wild in about 700,000 ha of southwest Morocco and also in the Atlas and Anti-Atlas mountain ranges. The Berber people have a long history of trading Argania oil, producing wildcrafted Argania oil, Argan wood and honey in the Argan forests. The tree is long lived, has exceptionally hard wood, and olive-sized fruit which fall after they turn yellow. Inside each fruit is a seed containing from one to four white, bitter, oil-rich kernels. It is from these kernels Argania oil is pressed. Argan trees can survive prolonged heat, moderate frosts, and prolonged drought. The tree is also a source of forage for sheep, goats, and cattle. The timber is used for carving, handles for tools, poles, and as fuel. The tree, which was unsustainably harvested for timber and fuel in the 20th century, is now protected in Morocco and has been replanted in areas which were clear-cut fifty years ago. These new plantings of Argan trees are interspersed with Caper bushes.

In Israel, the meager yield of nuts from wild Argan trees of 8 kilos per tree per annum has been increased to yields of 70 to 100 kilos in domesticated strains. These higher yielding strains are now grown in orchard formats. The Argan tree has also been introduced to Spain and the Canary Islands for reforestation. In irrigated format, the trees produce nuts twice a year but during long periods without water, an Argan tree may not produce nuts at all.

The nuts fall off the branch when they are ripe. Usually, they are dried in the sun on the ground. The outer pulp contains approximately 20% sugars, 13% cellulose, 6% protein, and 3% fat. The kernels are separated from the nuts after the nuts are peeled of pulp. Traditionally, the nuts are then roasted and ground to a paste. Tepid water is added to the paste and the oil decanted as it floats to the surface. In more modern methods the nut is cleaned of the dried pulp by being squeezed between rollers. Then the nuts are cracked mechanically. The oil-bearing kernels are separated by a centrifuge. The kernels are cold-pressed after roasting. The first press produces the finest oil and the second pressing yields a slightly inferior grade which is used for cooking. Argania oil is also used in cosmetics and is very good for the skin and hair. The oil is especially soothing on burns and stabilizes lipid metabolism in people who suffer from high levels of cholesterol. In its native range, it is made into two local foods. The oil is mixed with ground almonds and honey to produce the nut-butter "Amalou." Argania oil is also mixed with wheat germ and honey into a breakfast gruel called "Sematar."

The major fatty acids in Argania oil are: Myristic 4.3%, Palmitic 13.5% to 14%, Linolenic 4.6%, Steracic 5.6% to 5.7%, Oleic 45.2% to 47%, Linoleic 31.5%.

Oils For the Future

Here are some suggested oils for the future. These are all underutilized oils, some which can be harvested from wild trees today. These oils will become more important as agricultural systems move toward sustainability and diverse plants come into a wider continuum of human usage.

Baobab

This huge, water-storing, African tree (*Adansonia digitata*), is often the only island of shade in its native landscape of dry grass and brush. Baobab trees produce a large, egg-shaped, and down-covered fruit which is full of oil-rich seeds imbedded in white pulp. The pulp is rich in vitamin C and has a sharp, pleasant taste. It can be eaten or dried to make a lemonade-like drink. The pulp is sometimes added to porridge or used to curdle milk for cheese-making. The seeds themselves are the source of Baobab oil. Once separated from the pulp, they are usually fermented, pounded and crushed into a paste from which the oil is extracted by pressing. Sometimes the seeds are fried, ground, fermented, and formed into butter-like balls which are used to flavor soups and stews.

These are much localized uses of the Baobab seeds, which are underutilized relative to their oil producing potential. Very little research has been done on the composition of Baobab oil except in relation to its cosmetic properties. The oil is a semi-fluid, golden oil which contains nearly equal amounts of saturated and monounsaturated and poly-unsaturated fatty acids. The saturated acids are Palmitic 18% to 30%, Stearic 2.0% to 9.0%, and Arachidic 2.0%. The unsaturated acids are Oleic 30% to 42%, Linoleic 20% to 35%, and Alpha-linoleic 1.0% to 3.0%.

Balanites

Balanites trees (*Balanites aegyptica*) are long-lived desert trees. They are spiny and green-trunked with succulent green leaves. They live in wadis and plains with deep sandy soils and tap into shallow ground water or drink up floodwaters which seasonally water the sands. They are found in these types of areas all over the Middle East and North Africa. They are not tolerant of frost. The trees bear generously producing fruits which outwardly resemble dates. However, the pulp is sticky and bitter. The internal seed is very large. The fat content of the seed is high; 40% to 48% of it is edible oil. The kernels are freed from the pulp by sawing the fruit in half. The hard pulp and shell can be used as charcoal. Sometimes, the pulp is separated for making sweets and fermenting to alcohol. The sap of the tree is full of Saponins and used as an herbal remedy for Schistosomiasis, also known as, Bilharzia. The oilcake remaining after pressing has 36.8% protein.

Balanites oil is sometimes called Zycum oil after the Arabic name for the tree. Balanites oil is stable, golden in color, and prized for cooking. The major fatty acids in Balanites oil are: Oleic acid 31%, Linoleic acid 43% to 45%, assorted saturated acids 24%.

Balanites trees, which grow in great stands in poor, arid countries like Sudan, are underutilized. They can be a new sustainably-managed source of edible oil and other products, which would greatly benefit the population of these very arid areas.

Marula

The Marula tree (*Sclerocarya birrea*) is related to the Mango and the Cashew. It is a huge, wild tree with a light gray trunk and bluish leaves. The Marula tree towers over the African bush. They fruit copiously and drop within a few weeks giving a year's crop of juicy yellow fruits with large pits. The fruits are devoured by wild animals, gathered by villagers, eaten fresh, made into beer or brandy-like liquors, dried into fruit leather, and cooked into marmalades. The large seeds can be pressed into a superior cosmetic and cooking oil. Each pit contains two small nuts which are round and resemble hazelnuts. Placing a Marula nut on the coals of a fire makes the fibrous exocarp brittle and easier to crack. Once cracked out of the shell, the nuts can be cold pressed for a light, yellow, nutty-smelling oil with a large proportion of monounsaturated acids and natural antioxidants which make the oil very stable. The oil is used on the skin to heal and to cleanse, in food as a cooking oil, to treat leather, and to preserve meat. The major fatty acids in Marula oil are: Oleic 70% to 78%, Linoleic 4.0% to 7.0%, Alpha-linoleic 0.1% to

0.7%, Palmitic 9% to 12%, Stearic 5% to 8%, Arachidonic 0.3% to 0.7%.

Tallow Nut

The Borneo Tallow nut tree (*Shorea stenoptera*) grows wild in the tropical rainforests of South East Asia, Indonesia, and Borneo. The tree grows in fresh water swamps, marshes, along river banks, in shallow peaty soils, and alluvial soils. This large tree bears a generous crop of egg-shaped, winged, woody fruit about 4cm long and brown or black in color. The fat content of the fruit varies from 45 to 70%. The oil is a substitute for cocoa butter. It is also used in soap, cosmetics, candles, and medicines. Fruits are gathered as they fall from the tree or they are trapped in bamboo baskets as they float downstream. The fruits must be dried to prevent germination. If left in wet or damp conditions, they will sprout. One method of extracting the oil depends on the seed spouting to split the hard shell, then removing the kernels by hand and sun drying them before pressing to extract the oil. Another method is that of kiln drying, which loosens the shell. When the kernels are removed they can be cold pressed or pressed traditionally between two hardwood boards. The major fatty acids in Tallow nut oil are: Palmitic 18%, Stearic, 43% to 44%, Arachidic 1.1%, Oleic 37% to 38%, Linoleic 0.2%

Mowrah Butter

Mowrah butter also called Illipe butter comes from two related species of trees from India, *Maduca longifolia* and *Maduca latifolia.* (The binomial names of these trees are also spelled *Madhuca longifolia* and *Madhuca latifolia.*) These trees have very different needs and growth patterns, one is quite tropical in habit and found in monsoon forests while the other grows in Northern India's montane forests and is tolerant of frost. Both trees are drought tolerant and grow in marginal areas unsuitable for farming. Wildcrafted trees bear 10 to 45 kg of fleshy fruits between 2.5 and 5 cm long. Each contains 1-4 seeds. These seeds contain 55% to 60% fat. The kernels are not edible and neither is unrefined Mowrah butter which is sometimes used for candles and soap, also to prepare wool for spinning. The refined oil is used very much like other edible oils, in cooking and making vegetable ghee.

Owala Butter

Owala butter is a solid fat extracted from the seeds of the *Pentaclethra macrophylla*, a West African tree. The fat is considered suitable for cooking and can be fermented to produce a strongly flavored condiment for soups and stews. The potential usefulness of this tree is being investigated by the Food and Agriculture Organization of the United Nations (FAO) and the World Agroforestry Centre (ICRAF).

Kange Butter

Allanblackia oleifera seeds are processed locally in Central Africa into kange butter and *Allanblackia stuhlmannii* is processed into a similar substance called makani fat. Both products are used locally. Trees of the Allanblackia family bear large fruits up to 12 inches long with approximate 50 seeds in each. The seed kernels amount to 60% to 80% of the seed's weight. Grinding and pressing the seeds produces an oil which quickly solidifies. These substances are unusual in the fact they are almost totally composed of oleic and stearic acids. Currently these fats are used in soapmaking but further refining produces an edible butter.

Dika Butter

An edible fat called dika butter is made from the seeds of the *Irvingia gabonensis*. These seeds can also be ground into flour. Little is known about dika butter except it is also considered a soothing cosmetic oil as well as an edible one.

Brazil Nuts/Paradise Nuts

Brazil nut trees (*Bertholletia excelsa*) and Paradise nut trees (*Lecythis ollaria*) are found in the Amazon basin on well drained alluvial soils. Unfortunately these huge trees with their excellent timber are often cut first when virgin forest is cleared and wild-crafted material is not as abundant as it once was. This is also true of the Paradise nuts which seem to be a larger subspecies of the Brazil nut. Nevertheless nuts are still being harvested from approximately 500,000 trees. The nuts are contained in pods 1.5 to 2 kilos in weight containing between 12 and 20 nuts. Understandably, Brazil nuts are not gathered on windy days because of the danger of being hit by a falling seed pod from a tree 50 meters in height, an accident which can be fatal. Most trees produce 100 to 300 fruits per year but trees tend to alternate bearing.

Plantations of Brazil nut trees are rarely successful because of the long period between germination and nut bearing, a minimum of 12 years and an equally long period until the tree bears enough nuts to be profitable (24-30 years). This is one of the few economically important plants which are exclusively harvested in their natural rain forest habitat. As it turns out, Brazil nut flowers require a certain species of bee for pollination! The bees in turn require a certain species of orchid to survive. Male bees must acquire the fragrance of this particular orchid in order to attract female bees. If the forest is damaged by clearcutting, the orchids disappear, along with the bees and the Brazil nuts.

Most Brazil nuts are sold for export as whole nuts. However, broken or damaged nuts are converted to oil which is very much a secondary product of the nut collecting industry. The pods have to be dried, and then cut open with a machete-like knife. Nuts are soaked for 24 hours, boiled for five minutes, and then hand cracked. After cracking the nuts have to be dried and kept cool to avoid rancidity. The oil is used much as butter would be, in baking, cooking, and confectionary. Sometimes the Brazil nut is called the "butternut," which should not be confused with the nickname of a wild North American forest tree with the same nickname.

The major fatty acids in Brazil nut oil are: Myristic 0.6%, Palmitic, 15.4%, Stearic 6.2%, Oleic 48% to 49%, Linoleic 29.8%.

Babassu Palm

The Babassu palm (*Orbignya martiana* also *Orbignya oleifera*) is native to areas with high temperatures, fertile soils, and generous rainfall. Oil is pressed from the 1-4 bunches of seeds produced each year by Babassu palms. The individual fruits are between 8 and 15cm long. A bunch may be composed of many hundred of individual fruits and weigh 90 kilograms. The fruit contains between 3 and 8 kernels with an oil content of 60% to 70%. The fruit has several layers including a hard outer shell. These shells must be cracked by using a club or an ax or by 1 ton of pressure from a commercial cracking machine. The shells are used as charcoal and fuel. The next layer, the mesocarp, is ground for animal feed. However, the oil rich kernels inside the mesocarp are roasted and mashed. The oil is floated off by hot water. It is similar to coconut oil, mild in flavor, colorless, and does not easily go rancid. It is currently raised for margarine production, but also as a salad oil and for cooking and food processing. Oil cake is added to cattle feed. The main fatty acids in Babassu oil are: Caprylic 4% to 8%, Capric 6% to 7%,

Lauric 44% to 46%, Myristic 15% to 20%, Palmitic 6% to 8%, Stearic 3% to 5%, Oleic 12% to 18%, Linoleic 1.4% to 2.8%.

Caryocar Oil

Caryocar oil is produced from up to 15 related species of rainforest trees found in the Amazon Basin. The trees are mango like in appearance with broad leaves, large yellow blossoms and large round fruits containing edible kernels and layers of oil-like fat. It is this fatty layer which is extracted and used for cooking. The kernels are difficult to crack but also contain an oil which is even more prized by locals than the oil from the fatty layer. Kernels which are not eaten are usually roasted lightly and pressed. Processing of both oils should be done quickly after harvesting the fruit to prevent rancidity. Processed oil is kept in cool places in tightly closed bottles until use. The major fatty acids in caryocar oil (from the mesocarp) are: Myristic 1.5%,Palmitic 41%, Stearic 0.8%, Oleic 53.9% and Linolenic 2.6%. The major fatty acids in Caryocar Kernel oil are: Myristic 1.5%. Palmitic 48.4, Stearic 0.9%, Oleic 53.9, Linolenic 3.3%.

The aforementioned oils all come from trees. This is an alien concept to many people. However, excellent oils can be obtained from a wide range of plants. Many are easier to cultivate and more thrifty of water and other resources than corn, soya, and rape seeds. Some such as grape seed are the unwanted by-products of a crop cultivated for other purposes. Some are common garden plants. Just to name a few possibilities: safflowers, poppies, grape seed, flax, pumpkin seeds, calabash seeds, musk melon seeds, mustard seeds, watermelon seeds, sunflowers, sesame seeds, oyster nuts, many, many bean types, a dozen grains, and many interesting semi-wild plants like buffalo gourds are easily found sources of edible oil. All produce excellent edible oil in commercial quantities which can be produced using simple and readily available technologies.

Why then, does the industrial agricultural business complex place their emphasis on soil-destroying, water-use intense crops such as corn, soy beans and rape seed?

Are they truly "easier" to cultivate?

Chapter 9
Cloth That Grows On Trees

In areas of the world which had no cotton plants, no fiber bearing animals like sheep, goats, or llamas, and no silkworms, a different kind of cloth was made. This special cloth is "barkcloth," the cloth that grows on trees.

Barkcloth is a non-woven fabric made in tropical and subtropical countries from the soft inner bark of certain trees. It has been made and used in parts of Africa and India, the Malay Peninsula, Samoa, the Hawaiian Islands and the Fiji Islands, and the Pacific Northwest. In Polynesia and parts of Central America, barkcloth, perhaps reached its highest perfection.

Lengths of branches of species with long fibered inner bark, primarily, trees of the Moraceae family, such as the fig (*Ficus carica*), the breadfruit (*Artocarpus altilis*), or the paper mulberry (Broussonetia papyrifera) are selected and cut off the source tree. Alternatively, young stems are cut from those trees which are of small diameter. The outer bark is removed. The inner bark is cut in narrow strips and then alternately soaked and beaten with a grooved or carved wooden mallet until the fibers are well matted and become thin and flexible sheets. These sheets of soft inner bark are pounded on wooden anvils to further soften and expand them.

Sometimes the sheets are joined together to make huge cloths used as carpets or ceremonial hangings. Smaller pieces, sometimes waterproofed, are used as clothing. The barkcloth is usually bleached in the sun, then worked by hand into even softer sheets which are then finished into a variety of items. Many texts which mention "paper" clothing in Polynesia are actually referring to barkcloth.

This barkcloth is neither spun nor woven but durable and beautiful nonetheless. The barkcloth can be processed to a paper thinness, felted together to make tapestry-like hangings, waterproofed with the sap of other plants or painted and stamped with designs. It is versatile and produced from renewable resources: trees planted especially for the purpose.

Barkcloth is made into clothing, rugs, wall hangings, tents, shrouds, shoes, and ceremonial costumes and ceremonial gifts. In tropical regions around the equator and on many islands, people still make clothes and ceremonial textiles from barkcloth.

Barkcloth can be decorated by free-hand painting, by stencilling, tie dying, or by rubbing it over carved pattern-blocks. The dyes used to decorate the cloth are derived from the other local plants which are available so color, style, and design vary greatly from place to place and island to island.

Barkcloth has been belatedly recognized as a supreme expression of folk art. To quote a 2006 UNESCO bulletin: "KAMPALA, Dec. 3 (Xinhua)—Uganda's barkcloth has been named as part of the world's collective heritage recognized by the United Nations Educational, Scientific and Cultural Organization (UNESCO)."

Augustine Omare Okurut, head of the Uganda National Commission for UNESCO, is quoted as saying, "The proclamation of the art of barkcloth making in Uganda as a masterpiece of the world's intangible heritage is an honor to Uganda and a recognition of the indigenous textile production skills of Uganda craftsmen."

Making Barkcloth in Uganda

Lately, the cloth which grows on trees has become big business in Uganda. The barkcloth described below is made by the Baganda people of Uganda (and called "masaka"), from the inner bark of the mutuba tree

Ficus natalensis
©iStockphogo.com/Christophe Cerisier

A skilled barkcloth producer always carves his own "merit," the wooden mallet with which he beats the bark. With this merit he beats the bark for several hours, long enough to give the cloth the right thickness and size. The sound of the beating is unique and can be heard over long distances

Another grade of barkcloth is made from the long fibered outer bark of the Ficus natalensis or Natal fig. The farmer uses a natural "knife" made from the stem of a banana leaf to remove the living bark. This way he will not damage essential parts of the tree.

After the bark is harvested, the tree is wrapped in banana leaves to protect it from the sun. This way the tree immediately starts developing new bark. The better the tree is taken care of, the higher the quality of the material and the more quickly the tree recovers.

Bark cloth from Mali
©iStockphogo.com/Roberta Bianchi

The *Ficus natalensis* tree can be stripped once a year over a period of at least 40 years. This makes barkcloth a very sustainable and environmentally friendly material, as compared to cutting a tree for timber (which requires a thirty year investment in growth.)

Another quote about barkcloth, this time from the news bulletin of Art Nature Design (AND):

"AND (www.ArtNatureDesign.nl) is a company which has a special interest in connecting "poverty reduction" to "environmental issues" in third world countries. Through the "African skin" project European designers developed new products out of the sustainable material 'barkcloth' for the international market. Essential to these designs is the production has to take place in Uganda to generate income for Ugandan people and to improve local environmental aspects at the same time.

"In Uganda, for centuries local farmers manufacture a beautiful textile like material called 'Barkcloth.' The most remarkable thing about this material is the tree from which the material comes has the property to withstand complete removal of its bark. This aspect makes barkcloth an extremely environment friendly material. The value of the material barkcloth on the European market is high.

"AND has a business partner in Uganda which is capable of producing high standard quality products. AND sends designers to Uganda to teach the women how to produce office supplies, design bags, and gifts. AND also sends machinery to this company to increase production and quality."

Barkcloth in Samoa

Barkcloth has, of course, as many names as it has colors and styles. In Samoa, one of the centers for this craft, the cloth is called "siapo."

Manufacturing barkcloth in Samoa is a very ancient craft, probably one of the oldest ways of making cloth, and it has been practiced on Samoa and surrounding islands and refined for thousands of years. Hence the tapa cloth of the Pacific Islands is considered the most advanced and varied of barkcloth art and is perhaps the most colorful.

A Tongan tapa artist and her cloth
©iStockphogo.com/gprentice

In Samoa, tapa cloth, or "siapo," as it is known in the Samoan language, is still considered a major art form and an important symbol of Samoan culture. Tapa cloths have been used for many purposes throughout Samoan history, from regular traditional clothing to burial shrouds, and ceremonial items. Tapa cloth items are still commonly used in Samoa for traditional purposes as well as for everyday use in such ordinary items as bed coverings, wall hangings, room dividers, and household decorations. Other items are produced especially for the tourist trade.

Tapa from Samoa
©iStockphogo.com/Victor Ioramo

The bark used to make tapa cloth comes primarily from the mulberry family (*Moraceae*). The tree stalk is carefully pruned and tied so it grows straight with very few lateral branches. The tree is harvested once it reaches one-to-two inches in diameter and has an unblemished surface. The bark is stripped off the stalk in one piece. From the stripped bark the softest inner layer, called the bast, is separated from the tough outer layer which is preserved and used for tinder.

The inner layer, or bast, is scraped clean with an edge of a clamshell to clean away any remnants of the outer bark. The scraping also softens and spreads the wet fibers. The narrow strip of bast is then placed on a wooden anvil and pounded by a wooden beater while more water is added to keep the fibers from drying out. This beating process causes the bast to flatten and spread to become wide, thin pieces of cloth. This rather stiff, unfinished cloth, called u'a, is weighed down with stones so it cannot curl to be dried and bleached by the sun.

To make cloths of various sizes, strips of u'a are pasted together with arrowroot paste, called masoa. Holes in the cloth are patched with u'a and masoa taking care the edges of the patch are flush with the edges of the imperfection. The cloth is dried and made flexible by handling and working with the fingers. Once the cloth is both dry and flexible, it is ready to be decorated.

The traditional design elements used in siapo decoration are plant or animal motifs, sea based designs, or other images from Samoan life. Common examples include shells, starfish, pandanus leaves, frond patterns, and wave patterns. Some traditional motifs can be traced back to the beginnings of the Polynesian migration in Asia. Other patterns are much more recent, such as the use of lettering incorporated into the designs to spell out names, events, or dates. Design motifs are typically presented within a grid created by rectangular or oblong sections of color.

Three techniques are used in creating designs: the rubbing method, the freehand method and a method resembling tie dying.

The rubbing method uses a design board to imprint designs on the bark cloth. An unfinished cloth is placed on a design board which has been covered with dye. The top surface is then rubbed to transfer the design from the board to the cloth.

In the freehand method, each design image is hand painted on the surface of the cloth. These freehand designs are sometimes embedded in geometric shapes of contrasting color.

With the third method, wax or oil is used to prevent dye transfer to well defined areas of the cloth as the cloth is rubbed over a flat board covered with dye.

The dyes used in decorating tapa cloth are derived from plant sources. O'a, which is extracted from the bark of the blood tree (*Bishofia javanica*), is a brown dye which is the base for all other dyes. As it ages, it darkens from a pale beige to a dark rich brown. Black dye, or lama, is made by burning the nut of the candlenut tree (*Aleurites moluccana*), a particularly beautiful local tree with maple-like leaves. The charcoal of the nut is collected and mixed with o'a to make lama. Loa is a bright red dye made from seeds of the lipstick tree (*Bixa orellana*) mixed with o'a. Yellow dye, ago, is extracted from turmeric roots (*Curcuma longa*) which look like small soft carrots. This strong yellow dye is a popular color but often fades from older cloths to a pale gold. Purple coloring, soa'a, comes from the sap of the banana plant (*Musa sapientum*). This dye is no longer used because collecting its sap for the dye's manufacture does too much damage to the banana plant and the ceremonial uses of this odd, twilight colored, cloth have become unfashionable.

Barkcloth in Fiji

Fijian barkcloth, "masi," in the Fijian Islands is the only barkcloth which is made exclusively from the bark of the paper mulberry tree (*Broussonetia papyrifera*). Other suitable raw materials exist but are not generally used. Masi has many forms, thicknesses and design styles. At the time of the first European contact with Fiji, by far, the greater amount of masi was plain, finely made, bleached cloth used for clothing and household items.

The majority of masi pieces in museum collections is patterned and colored in a variety of ways. There are many styles appropriate to different island groups, some influenced by population movements or trade from other communities such as Tonga and Samoa, which also have long traditions of highly colored and skillfully dyed bark cloth.

Masi is normally made and decorated by women, except in the highlands of Viti Levu, the largest island, where the decoration is undertaken solely by men.

A distinctive way of decorating found only on Fiji uses stencils which were traditionally made of banana or pandanus leaf, but more recently from exposed x-ray film, cardboard, plastic sheets or other appropriate material. The most usual colours are black and red or dark brown. The design is built up from the border, in which a limited number of motifs is used repeatedly making a well planned pattern. This kind of cloth is called "masi kesa."

Bark cloth from Fiji
©iStockphogo.com/elianadulins

The most popular style is associated with the Cakaudrove District of south-eastern Vanua Levu and the neighboring large island of Taveuni. The cloth is crisply and repeatedly folded, the edges being marked with black dye. Once unfolded, the outlined panels are infilled with dye using different colors and designs. Edging bands of stencilled decoration are also incorporated into the design and the result is unusually striking and distinctive.

Masi has always been an important trade item between Fijian communities, not all of whom made their own, but may have specialized in other goods such as a fine matting or pottery, or elaborately carved wood. This type of trade also took place between Fijian islands and Tonga and Samoa. Vast amounts of masi were produced for ceremonial gift exchange between chieftain families, as well as, for inter-island trade, both within the Fijian islands and throughout western Polynesia. More recently, as with other Pacific island groups, the needs of the tourist trade for items made of this traditional commodity have become pre-eminent, influencing form and designs.

Barkcloth in the North American, Pacific North West

Cedar bark textile was used by indigenous people in the Pacific Northwest region of modern-day Canada and the United States. Historically, most items of their clothing were made of this type of barkcloth. The name is a confusing misnomer, as cedar bark textile is made from material taken from Western Redcedar (Thuja pilatica) and Yellow Cypress (*Callitropsis nootkatensis*) bark, not true Cedar bark, as true cedars are actually Old World trees, not naturally found in North America.

The red cedar tree
©iStockphogo.com/Erik Odegard

After bark from a Western Redcedar or from a Yellow Cypress was peeled in long strips from the trees, the outer layer was split away, and the flexible inner layer was shredded and processed by the wetting and pounding of their long fibers. The resulting felted strips of bark were soft and could be plaited, sewn, or woven into a variety of fabrics which were either dense and watertight, or soft and comfortable depending on the length of the pounding process and what had been added during the processing. One way to make the barkcloth waterproof was to add pine gum during the processing of the cloth. Sometimes, to soften the fibers, the bark strips would be soaked in sea water.

Women wore skirts and capes of soft redcedar barkcloth, sometime adorned with feathers while men wore long capes of barkcloth into which some mountain goat wool was woven for warmth and decorative effect. Bedding and wall hangings of barkcloth were not unknown in the Pacific Northwest, neither were sacks and bags of this material. Still the primary use for barkcloth in the Pacific Northwest was for articles of clothing.

Barkcloth in New Guinea

Barkcloth is made in New Guinea for ceremonial costumes and exchanges as well as for sale. Designs are specific to clans. Individuals may reinterpret traditional designs or create new ones. The rights to a design are often owned.

Since Barkcloth making was widespread throughout Polynesia and parts of Oceania the craft was spread by contact between islanders, probably reaching New Guinea over a thousand years ago. In New Guinea, the art of making barkcloth hats and ceremonial costumes reached its peak, a fact which belies the widespread anthropological idea that New Guinea natives were on the very end of the cultural telephone, primitive and slow to adapt. Now store cloth has replaced barkcloth for daily use, but it continues to be made for ceremonies and for sale.

Most barkcloth made for sale comes from Oro Province. Usually they are geometric patterned pieces which are sold in 1-2 yard (1-2 m) squares, framed as paintings or inset into coffee table tops, folio folders and other decorative items.

A barkcloth pattern from the Asmat region of New Guinea
©iStockphogo.com/Øystein Lund Andersen

Cultivated paper mulberry (*Broussonetia papyrifera*) is the preferred bark, although breadfruit (*Artocarpus altilis*) and other forest trees are used if they have suitable thick, fibrous inner barks. In New Guinea, damp pieces of bark are soaked in fresh water or sea water, over-lapped, and beaten together to form large sheets. Sheets are folded and beaten out, refolded and beaten out yet again and again to make a uniform cloth without holes. Early explorers in New Guinea wrote about villages resounding with groups of chanting women beating out barkcloth and about the natural clays and minerals formulating the dyes with which they were decorated.

New Guinea Tapa can be made as thin and fine as lace or layered into lengths with the consistency of thick felt. Plain tapa, sometimes bleached to pure white, was often important in traditional ceremonies, but it was seldom interesting enough to be collected by outsiders. Pattern books of tapa made in the colonial period were very popular. Patterned tapa is still the choice for commercial sale.

Patterns, whether traditional or contemporary, add meaning to barkcloth beyond decoration. Alfred Gell writes in *Wrapping in Images: Tattooing in Polynesia* (Oxford University Press, 1993), both barkcloth and tattoo designs are seen as an additional layer of skin wrapped around the individual.

Tapa patterns are created elsewhere by staining, painting, stamping, and stenciling. However, in New Guinea, the designs are exclusively hand painted. Traditional colors are somewhat different as well as they come from locally available clays, minerals, and plants, as well as, charcoal.

Barkcloth in Japan

In Japan a totally different tradition and method of making barkcloth emerged from the ancient Ainu culture. Elm or linden trees are the source of the thread for the Ainu's woven textile.

The Ainu, who are distantly related to the Australian aborigines, are not oriental but archaic Caucasians. How and when they arrived on the four Islands of Japan is unknown but they have apparently been there for a very long time.

They produce one of the finest and most astonishing barkcloths in the world. It is not a pounded or felted substance but one which is actually woven from a yarn which is made from the fibers in elm bark or linden bark sometimes blended with the fibers of the common thistle. This strange textile has to be seen and touched to be believed. Needless to say the process was very labor intensive.

Elm trees

©iStockphogo.com/Stephen Shockley

"Attus," the garments made of this odd cloth are usually made from the bast fibers of a Japanese elm tree (*Ulmus davidiana var. japonica*) which is native to Hokkaido. First the tree is cut down and the bark is removed. The fibers, taken from the inner layers of the bark, are soaked in water to soften, bleached in the sun, and then split into fine, fibrous strands. The strands are joined together into thread and this thread is woven into cloth. The finished product is a thick, stiff cloth of a brownish color, like the bark fibers from which it is produced. This cloth is then sewn into an attus garment. The cloth itself is called "attush."

Attus which were worn for everyday wear, did not have much decoration, but those made for ceremonial wear, were decorated on the back and around the sleeves with patterns in navy and black applique. This unique pattern " ," like a parentheses on its side, is repeatedly embroidered on the appliqué.

Royal Ainu dress
©iStockphogo.com/Alexander Gatsenko

This pattern is worn to keep away evil spirits and bad luck. It is found not only on attus but also on dark blue cotton garments with similar patterns, and on garments with white applique on dark cotton, called "kapara amip."

The applique and embroidery work was always done by women. From mother to daughter, this unique tradition was handed down from generation to generation. By making these traditional garments such as the attus, Ainu women not only clothed their families, but in their eyes, actively protected them from evil as well.

Basho-fu: Cloth from Banana trees, a unique Okinawan art form

Thread banana trees

©iStockphogo.com/luxxtek

Banana fiber cloth is made of fibers of a special species of banana called a "thread banana" plant (*Musa balbisiana*). Though similar to the plants which produce edible bananas, "basho" (or more specifically, "ito basho," or "thread banana" plants) are remarkable for their fibrous stems instead of for their fruit. Basho plants grow into "trees" of about two meters in height. (Banana species include some of the largest herbaceous plants in the world, most large enough to be considered trees.) The fibers in the "trunks" of basho plants basho can be split into fine strands, tied together into thread, and woven into cloth. This cloth is called "basho-fu."

These trees can be found in numerous locations across the Ryukyu islands and the cloth made from them was used as tribute payments to the Okinawan royalty. In those days, basho-fu was worn by everyone from kings to commoners. Basho-fu has long been favored for the crafting of summer kimonos because of its airiness and smooth, crisp surface. Like linen, hemp, ramie, and other long vegetable fibers, basho-fu does not stick to the skin in hot weather, making it perfectly suited to the hot Okinawan climate.

In the old days, bolts of plain-colored, striped and kasuri (ikat) basho-fu were woven in many villages. Nowadays, however basho-fu is a luxury cloth which is made only in the village of Kijoka, on the island of Okinawa. Making basho fu is almost a lost art.

Cloth from Dead Sea Fruit

Surely the most exceptional raw material for cloth making comes from Dead Sea fruit, the fruit of the huge tree-like desert herb *Calatrops procera.*

Its curving branches, large rounded leaves, striking purple and white flowers and huge elliptical fruit are common sites in arid and saline areas in the middle east and Africa. However it is very poisonous and apart from a few obscure medicinal properties of the milky sap the plant is considered useless and regarded as a noxious weed. "Dead Sea fruit" in English is an expression for something which is assumed to be of great value but turns out an illusion, to be dust and ashes.

Dead Sea fruit in an Israeli garden
Photo by the author

The big green lush looking fruit of the Dead Sea apple is not edible and filled with long white fibers and dry seeds which float away on the wind when the skin of the fruit dries and is broken. These dry snow white fibers were once collected and woven into cloth in Biblical times, probably for use of the priestly caste, in the strangest example yet found of the "cloth which grows on trees."

Chapter 10
Vitamin Trees

The stories of great sea voyages and of the exploration and the colonization of new lands are often stories about great hardship and suffering. The first colonists of European origin in North America, whether it was the English colonists in Jamestown and Roanoke or the Spanish colonists in Los Angeles, literally died of want, even though native people around them continued to thrive. Sailors on long sea voyages often perished from scurvy despite the fact they had adequate supplies of stored food. In the gold rush to Alaska that started in 1887 and continued throughout the 1890s—more people died from scurvy than actually found any significant amount of gold, even though veteran settlers and native Alaskans never suffered from this ailment.

Between 1910 and 1912 British explorer Robert Scott and the Norwegian explorer Roald Amundsen were locked in a race to be the first to arrive at the South Pole. Amundsen, who had planned his expedition carefully and packed extremely concentrated food supplies, brought his expedition to the South Pole and back safely. He depended on his high protein, high vitamin, and high fat rations to keep the expedition party healthy and depended on teams of sled dogs for fresh meat and transportation.

Scott reached the Pole 33 days after Amundsen. He had depended on Manchurian ponies and man-hauling sled power to get to the South Pole and had set up supply depots at various points to help him get there and then back to base camp. Unfortunately, the ponies were useless in the snow and had to be shot. The stored rations were lacking in the concentrated nutrition needed to support men doing the heavy physical work of hauling sleds over the ice fields and could not keep Scott and the other members of the expedition healthy and strong. (A British team, attempting to recreate Scotts' trek and using his rations, clothing, and methods of travel, lost so much weight on the journey the experiment was called off after they reached the pole. They were airlifted back to base camp.) Amazingly, the difference in the vitamin content of the foods used respectively by Amundsen and Scott may have made the difference between life and death.

Vitamins had not been discovered at the time, but Amundsen says quite clearly in his account of his successful journey to the Pole, *"The packing of tinned provisions is of enormous importance to a polar expedition: it is impossible to give too much attention to this part of the supplies. Any carelessness, any perfunctory packing on the part of the factory will as a rule lead to scurvy."*

In order to stave off scurvy, Amundsen added dried vegetables to the traditional food pemmican, generally a mixture of meat and fat. He said, *"The pemmican we took was essentially different from that which the former expeditions had used. Previously the pemmican had contained nothing but the desired mixture of dried meat and lard. Ours had besides vegetables and oatmeal, an addition that greatly improves the flavor and as far as we can judge makes it easier to digest."*

Further additions to the food for the expedition included canned fruits from California, whortleberry and cloud berry preserves, fruit syrups, and dried fruits. The expedition team was also supplied with fresh meat in the form of seals hunted on the Antarctic coast, and dog meat from culled sled dogs. Even the specially-made hardtack biscuits for the expedition were fortified with powdered milk and oatmeal. In his account of the voyage, Amundsen also mentions as good sledging food, dried fish, chocolate, malt, "middlings," and sugar.

Amundsen, who was not trained in nutrition or medicine, had learned from the native peoples of the Arctic that staying well fed and warm in such a frigid climate meant staying alive. He even bought 250 reindeer hides and had craftsmen copy Eskimo clothing, with three sets of anoraks and trousers for each member of the expedition of varying weights, waterproofed, and with the fur inside. No one from Amundsen's expedition suffered from scurvy and his entire expedition team returned safely from the South Pole.

In contrast, Scott's expedition to the South Pole ended in disaster. Scott's expedition members wore clothing of wool and canvas, although their boots and mitts were made of reindeer hide and beaver pelts. Scott's "man-haulers," who would have needed about 7,000 calories per day per man during the long arduous trek over the ice, were seriously underfed as well as vitamin deficient. His rations were mostly comprised of biscuit, traditional pemmican, cocoa, butter, sugar, and tea with no vitamin food at all. A ration for a day's haul was 450 grams of biscuits, 340 grams of pemmican, 85 grams of sugar, 57 grams of butter, 20 grams of tea, and 16 grams of cocoa. Scott's expedition died to the last man in their tent, 11 kilometers from a food depot suffering from unhealed wounds, frostbite, malnutrition, and trapped by bad weather.

The unhealed wounds, general weakness and loss of weight Scott reported in his diary were as likely to have been symptoms of vitamin C deficiency as they were to be consequences of the hardships they endured. He could not have known this, of course, as the prevailing medical opinion at that time was scurvy was caused by improperly preserved food. However, the lack of vitamins in Scott's rations probably made the difference between life and death. Certainly the vitamin content of the food of the starving settlers, the sick sailors and the dying prospectors was a critical factor.

Too late for the early colonists, sailors, and explorers, vitamins were only discovered in early 20th century after it became known scurvy, pellagra, and many other mysterious illnesses could be induced by the prolonged consumption of a limited diet. Vitamins are complex substances vitally important for metabolic functions. A severe lack of vitamins can be debilitating or fatal.

The diseases caused by vitamin deficiencies at their worst seemed as merciless as biblical plagues and the fact no one knew what caused them was terrifying. Besides the most damaging and most common deficiency diseases, such as scurvy, "beri-beri", and pellagra, there was also "rickets" which bowed the legs and curved the spines of generations of children and was caused by a deficiency of Vitamin D. The twentieth century was well underway before it was discovered rickets could be cured by sunlight and cod liver oil.

In modern times the worst deficiency disease may be the cluster of symptoms caused by a lack of Vitamin A, or the failure of proper immune system functioning which it is thought to be a result of a deficiency of Amygdalin, which some call Vitamin B_{17}. This latest theory is very controversial with proponents claiming the astounding increase in cancer rates around the world is caused by a deficiency of Amygdalin in the modern diet and opponents counterclaiming Amygdalin is a dangerous, potentially toxic substance which has no effect on cancer whatsoever.

As in most controversial research, the veracity of claim and counter-claim will only be clear as evidence accumulates and is analyzed objectively.

Discovering the Cause of Scurvy

Early clues to the existence of these substances were noted by a surgeon of the British Fleet, Dr. James Lind, in the mid 1700s. He was aware thousands of British sailors were crippled or killed by scurvy, which was then thought to be a contagious infection. However, he had heard of the story of a seaman dying of scurvy who was abandoned on a desolate Atlantic island. The deserted man ate the fresh spears of grass he found and regained his strength. He was picked up by a passing ship and returned to his home telling the tale of how he had survived by eating grass like a beast.

Dr. Lind began to treat his scurvy patients with greens, oranges, and lemons, assuming some necessary element was present in these foods which was absent from the navy's rations. Eventually despite ridicule and official opposition he managed to make fresh lemon and lime juice part of the official rations of the British navy, leading to the British sailors to be nicknamed "Limeys".

In 1905, a Dutch scientist, Professor Peklharing, found milk also contained an essential substance and mice supplied with plenty of grains and fats would die if they were not supplied with milk as well.

In 1906, an English biochemist, Sir Fredrick Hopkins, discovered certain food factors among the cereals were important to health. Investigating an epidemic of beri-beri, which affects both the nervous system and the digestive system, he came to the conclusion those whose diets consisted mostly of polished rice (white rice) were much more likely to have beri-beri than those who ate brown or unhulled rice. His conclusion was there were elements in the rice hulls which prevented beri-beri. This was confirmed by the work of Dutch scientist Christiaan Ejikman, who gave chickens beri-beri by feeding them polished rice and restored their health by returning them to a diet of brown rice.

The word "vitamine" was first used by a Polish scientist named Cashmir Funk to refer to a group of compounds considered vital for life. It was thought all these compounds contained a nitrogen based component called an "amine." When it was discovered not all of these important factors contained nitrogen and not all of them were amines, the final "e" was dropped and the word came into common usage. Soon many scientists were investigating vitamins and trying to isolate them, some with an eye to synthesis and others to simply understand where they were found and why they were important.

In 1915, Dr. Joseph Goldberger began to suspect pellagra was caused by the lack of a vital element in the diet of poor rural people in America's south. Pellagra had become an epidemic with more than 10,000 people dying that year and many thousands more showing the dreaded symptoms which include weakness, lethargy, tremors, dry skin, and the tell tale "butterfly rash." He attempted to give himself pellagra by exposing himself and several volunteers to the bodily fluids of victims. This did not lead to infection. Dr. Goldberger also attempted to "catch" the disease by staying in insane asylums, which were often crowded with apathetic, confused, and sometimes violently insane "pellegrans." It was there he noticed the inmates might be very ill, even dying of the disease but the staff never had it. Finally he induced pellagra in convict volunteers by keeping them on a diet of rice, fatback, grits, cornbread, and sorghum syrup, the common food of the poor people in the American south. This proved pellagra was a deficiency disease.

Dr. Goldberger then proceeded to cure pellagrans with fresh milk and meat and yeast extracts. He was never able to isolate the exact element which cured them. The B-complex vitamins and the dietary essential niacin or B_3 (without which the rest of the B-complex vitamins cannot function), were only identified in 1937. Vitamin D was soon discovered as the missing element in cases of rickets and the lack of Vitamin K was identified as a cause of clotting disorders.

The suffering caused by vitamin deficiency diseases has mostly been forgotten in the modern world. Scurvy causes loss of weight, shortness of breath, pallor, hemorrhages under the skin, and the loss of teeth and hair. The victims sometimes convulse or lapse into delirium and usually die from internal bleeding. Beri-beri starts with depression, fatigue, and stiffness of the lower limbs. It may end with swelling of the limbs, paralysis, and burning pain in the extremities. Pellegra's symptoms are manifested on the rough, discolored skin, in the patients' inflamed mouths, and terrifying neurological symptoms which may include twitching, tremors, and delusions. Extreme cases may literally "foam at the mouth" like victims of the rabies virus.

The terrible consequences of dietary deficiencies should not be forgotten nor should the means by which these diseases may be cured. Lemon juice saved the British sailors from scurvy. The unfortunate seaman abandoned on a deserted island was cured by eating grass. Sufferers from beri-beri sickened and died with their cures all around them in the fields and markets of their villages. Rickets crippled children in well to do families while the cure for their ills streamed outside the windows of their houses in the form of sunlight containing ultraviolet B light, which human skin cells use to produce vitamin D. The goldrushers to the chilly north sometimes suffered and died with sources of vitamin C all around them in the form of the spruce tips and pine tips, wild berries, and in the livers of game animals. If they had known how to utilize these food sources, they could have cured themselves with a few good meals.

Among the common trees of the world there are those which can supply vitamins in sufficient quantity and quality to prevent vitamin deficiencies. Knowing where to find these essential dietary elements may not be a matter of life and death just now, but it is very useful knowledge. Some tree products are so packed with nutrition, the trees which bear them should really be designated as "Vitamin trees."

Vitamin A is a vitamin necessary for sight. It also has influence on hormonal functions. The consequences of severe Vitamin A deficiency can be seen in areas where the population cannot grow enough green leafy vegetables, carrots, or yams and does not have access to fish or liver meats. The most common symptoms of Vitamin A deficiency are loss of color vision and night vision, blindness, susceptibility to disease, and skin eruptions. Liver, carrots, yams, peppers, and cabbage are common sources of Vitamin A. What is not generally known is mangoes (*Mangifera indica*), apricots (*Prunus armeniaca*), thorn-plums (*Ziziphus spp.*), sea-buckthorn berries (*Hippophae rhamnoides*), Prickly pear cactus pads and fruits (*Opuntia spp.*) are also good sources of Vitamin A. Another extremely rich source of pre-Vitamin A is the Moringa tree (*Moringa oleifera*), which also contains various B vitamins as well as Vitamins C, E, and K.

Vitamin C is present in most green vegetables but peppers, broccoli, peas, tomatoes and cauliflower are especially rich in this vitamin and in emergency situations even young spears of grass or the tips of spruce trees will do. The symptoms of Vitamin C deficiency have already been described. All citrus fruits such as lemons (*Citrus limon*), limes (*Citrus aurantifolia*), grapefruitss (*Citrus × paradisi*), and oranges (*Citrus × ?sinensis*) are rich in vitamin C. Other good fruit sources are guavas (*Psidium guajava*), kiwifruit (*Actinidia deliciosa*), marulas (*Sclerocarya birrea*), papayas (*Carica papaya*), and pineapples (*Ananas comosus*). The common potato (*Solanum tuberosum* is also a good source of vitamin C, a happy accident which kept scurvy at bay among the poor in the cool countries of Europe. The ancient Chinese navy fed its sailors dried fruit and spouted seeds to maintain good health. Edible sprouted seeds of all kinds contain as much C by weight as lemon and lime juice.

Vitamin E is actually a group of fat soluble compounds which influence the body in the same manner. A lack of Vitamin E may cause sterility, neurological problems, anemia, and the wasting of muscle tissue. The most common sources of Vitamin E in the diet come from oils. Dietary Vitamin E is usually obtained from wheat germ, unrefined corn oil, or fish such as salmon, tuna, and sardines. What is less well known is many products of trees and other perennials contain Vitamin E in sufficient quantity to replace the common oils and the fatty fish if those sources are not available. These include olives and olive oil (*Olea europaea* subsp.), blueberries (*Vaccinium spp.*), papayas (*Carica papaya*), hazelnuts *(Corylus avellana)*, kiwifruits (*Actinidia deliciosa*), pine nuts (*Pinus pinea* and other *Pinus spp.*), mangoes (*Mangifera indica*), and almonds (*Prunus dulcis,* syn. *Prunus amygdalus*). Most nut oils, such as walnut oil (genus *Juglans*), black walnut oil (*Juglans nigra*), pine nut oil (*Pinus spp.*), argan oil (*Argania spinosa*), and macadamia nut oil (*Macadamia integrifolia* and *Macadamia tetraphylla*) are high in Vitamin E, also.

Vitamin K denotes a group of lipophilic, hydrophobic vitamins which are needed for the posttranslational modification of certain proteins, mostly required for blood coagulation, but also a number of other proteins which chelate calcium ions and are involved in bone and other tissue metabolism. The lack of vitamin K may cause clotting disorders including uncontrolled bleeding from the nose, bleeding gums, debilitating menstrual periods, internal hemorrhages, easy bruising, and wounds which do not heal. Lack of vitamin K can also cause brittle bones and lead to stress fractures. Green leafy vegetables are the best source of Vitamin K and luckily most places on the planet have vitamin K rich edible plants among the domesticated crops or the local wild plants. However, Vitamin K deficiency can appear in people who live in climates with long or severe winters because leafy fresh foods are then not available. The traditional remedy for a lack of fresh green foods is to add to the diet pickled and lacto-fermented foods such as kimche, sauerkraut, pickled cucumbers, or other preserved green foods such as kale or Brussels sprouts put up in brine. From the tree crops comes one of the best sources of Vitamin K and the easiest to preserve. Plums are excellent sources of Vitamin K. Oddly enough both lacto-fermented green food and dried plums contain more Vitamin K than do their fresh counterparts. Other good sources of Vitamin K can be found in avocados (Persea americana), tangerines (*Citrus × tangerina*), kumquats (*Citrus japonica*), kiwifruits (*Actinidia deliciosa*), cherry plums (*Prunus cerasifera*), and plums (*Prunus spp.*) (which are also known as "gages" in some areas).

The B vitamins are a crucial cluster of eight related water soluble vitamins which are very important to cell metabolism. They were originally thought to be one substance, which became known as Vitamin B but later it became apparent the original B was actually eight different vitamins with niacin or B_3 as an "enabler." A lack of any one of these vitamins can have serious consequences but a lack of Vitamin B_3, as in the case of the pellagra outbreaks in Europe and the US, was often deadly. Since pellagra only appeared where maize had been adopted as a staple crop it was once assumed there was something in the corn which caused the disease. Pellagra was sometimes called "the corn infection" and identified by the "Three D's," dermatitis, diarrhea, and dementia.

After Dr. Joseph Goldberger proved the connection between pellagra and diet it was asked why there was no pellagra in the predominantly corn eating societies in Mexico and South America. This was due to the traditional milling of maize foods with lime, or nixtamalization, which made the niacin in the corn available to the human metabolism. It was only when corn was adopted as a staple crop, but the nixtamaliztion method of preparation was left behind, that the disease appeared.

Good sources of B_3 are chicken, fish, beef, carrots, milk, and brewer's yeast which became the standard treatment for pellagrans after Dr. Goldberger used it to cure his patients. Other good sources of niacin are dates, avocados, mangoes, pears, and most tree nuts. Oddly enough there was a high niacin food available in the American south: the humble peanut. "Goober eaters" were almost never pellagrans nor were the people who habitually ate wild game. Regrettably, the pieces of this puzzle were not put completely together until the 1950's when it discovered why the niacin in corn was not available to some corn eating populations.

The other B vitamins are:

- B_1 or Thiamine
- B_2 or Riboflavin
- B_5 or Pantothenic acid
- B_6 or Pyridoxine
- B_7 or Biotin
- B_9 or Folic acid
- B_{12} or Cobalamin

The B vitamins together support the entire human metabolism, maintain skin and muscle tone, promote cell growth, prevent anemia, enhance the immune system, and help prevent disease. Most B vitamins have to be replenished regularly, the exception being vitamin B_{12} which is stored in the liver.

The lack of B_1 or thiamine deficiency is the culprit in outbreaks of beri-beri. This disease affected populations whose staple food was polished rice. Soybean products, liver, whole grains (including unhulled rice), whole potatoes, dairy products and "Quorn" (A meat substitute mainly marketed in Europe. The mycoprotein used to produce Quorn is extracted from a fungus, Fusarium venenatum, which is grown in large vats.) are all excellent sources of B_1. Almost all tree nuts, as well as, oranges and most citrus fruits are very rich in B_1. It is sadly ironic both pellagra and beri-beri, two of the worst of the deficiency diseases, are caused by mishandling an inherently nutritious staple food. In the case of maize, the maize is not treated to make its vitamins available to humans who eat it. In the case of rice, the rice is polished to extend its storage life, and the process greatly decreases its food value.

Vitamin B_2, riboflavin, is essential for the health of the skin and mucus membranes. It is relatively common in whole foods such as dairy products, whole cereal products, broccoli, wheat bran, and leafy green vegetables. It is found in tree nuts and the seeds of leguminous plants including Acacias (*Acacia spp.*), Moringas, Mesquites *(Prosopis spp.*), Carob (*Ceratonia siliqua*), and most common beans.

Vitamin B_6, pyridoxine, is necessary for healthy blood, circulation, and fluid balance in the human body. A lack of B_6 can cause hypertension, water retention, and anemia. Luckily there are many good sources of Vitamin B_6 including sunflower seeds, spinach, oatmeal, fish, beef, pork, chicken, wheat bran, seafood, peanut butter, soybeans, and lima beans. In the tree crops the bananas, plantains, and walnuts are the richest sources of B_6, followed closely by avocados, leguminous tree crops, other tree nuts, and mulberries. B_6 is found in mushrooms and edible fungi of all kinds including the bracket fungi and forest mushrooms which grow on living and dead trees.

B_7, also called biotin, is a commonly available vitamin. Deficiencies are rarely problematic in adults even though biotin plays a vital role in human biochemistry, especially in maintaining blood sugar and using the energy of carbohydrates. However, biotin deficiencies can be very serious in children and infants as a lack of biotin can delay development, stunt growth, and even cause mental retardation. Biotin is commonly found in red meat, seafood, eggs, mushrooms, soybeans, beans, cauliflower, milk, whole grains, breads, organ meats, molasses, and peanuts. In tree crops it is found in bananas (*Musa spp.*), chocolate (from the cocoa, *Theobroma cacao*), cashews (*Anacardium occidentale*), filberts (*Corylus maxima*), almonds, (*Prunus dulcis, syn. Prunus amygdalus*) and Brazil nuts (*Bertholletia excelsa*).

B_9, folic acid, is essential for manufacture of new cells and so is crucially important in the development of the human embryo. A lack of folic acid in a mother's diet can lead to serious birth defects. Luckily, it is found in useful amounts in almost all green leafy vegetables, most fish, and many root crops such as beets, turnips, parsnips, root celery, and also in liver and organ meats. B_9 is found in many fruits such as grapefruits, oranges, avocados, pomelos (*Citrus maxima* or *Citrus grandis*), and bananas, in berries such as mulberries, raspberries, strawberries and in tree vegetables like the leaves of the Moringa tree. Due to folic acid's importance in preventing birth defects and promoting healing many grain products, breads and cereals are "fortified" with folic acids.

B_{12}, a vitamin necessary for mental health, good memory and a healthy nervous system is one of the few vitamins not easily found in the plant kingdom. Good sources of B_{12} are red meat, milk, pork, fish, seafood, cheese, eggs, yogurt, and whole milk. Vegans and vegetarians are often deficient in Vitamin B_{12}. The vitamin is also found in yeast, beer, and malted foods with yeast being the richest source. There are no known reliable sources of Vitamin B_{12} among the tree crops and few in the plant kingdom. Because of this vegans and vegetarians must be very careful to seek out and eat Vitamin B rich foods.

In naming the sources of various vitamins a clear pattern emerges. There are some plants which are named again and again. The bananas and the plantains are powerhouses of nutrition. Tree nuts are excellent storable sources of much of what is needed for health. The avocado, unjustly criticized for its fat content, contains as many vitamins as a multivitamin pill. The leafy vegetables are vitamin powerhouses. The Moringa tree "towers" among them with its broad spectrum load of vitamins, making it a serious contender for the title "Vitamin Tree."

There are many things which do not grow on trees, but vitamins, with the exception of the elusive B_{12}, most definitely do.

Chapter 11
Trees and Their Names

It is very important, when working with trees, to learn the trees' Latin or scientific names. Many trees share a family name, but the behavior, habit, and/or requirements of the individual species may be very different. Because of this, it is important to identify positively the tree with which one is working and to be very sure research colleagues and other interested parties are referring to the same species and variety, and not some entirely different plant with similar common name.

The modern system for naming plants, usually known as "binomial nomenclature," was invented by Carolus Linnaeus, a Swedish botanist. He invented the system to end centuries of confusion among farmers, scholars, herbalists, and doctors about the plants they grew, studied, or used. This was a very important development in the science of botany as plants were known only by their local names, many very similar, before the Linnaean system came into use—and it was very difficult to describe or define exactly which plant was being referenced.

The Latin name or scientific name of a plant can now be used all over the world with generally accurate results. Though scientific names of plants sometimes change with new knowledge and new discoveries it is usually possible to find out the names and properties of specific species of trees and to avoid the kind of mix-ups which happen when different plants share the same common names.

To explain briefly how the system for naming living creatures works, it is best to start with the most commonly accepted five broadest categories of living creatures, the "Kingdoms."

These are:

Kingdom Monera: prokaryotic, microscopic creatures without membrane-bound nuclei or cell organs. Examples of such organisms are bacteria, cyanobacteria and spirochetes. Most are unicellular, but some organisms form chains and others have multi-celled phases. Nutrition is absorbed through cell walls. Prokaryotes generally have a single loop of DNA rather than chromosomes.

Kingdom Protista: eukaryotic organisms with membrane bound cell organelles, nuclei and chromosomes. Examples from this Kingdom include organisms which straddle the plant/animal divide such as amoebas and odd, difficult to classify, organisms such as giant kelps. Nutrition is by ingestion, absorption, or synthesis, so this is sort of a catch-all kingdom for organisms. It is difficult to categorize. Also some of the included organisms are motile.

Kingdom Fungi: non-motile organisms which form mycelia or hyphae and take their nutrition from dead and decaying material. Examples from Kingdom Fungi include molds, yeasts, mushrooms, blights, and mildews.

Kingdom Plantae: non-motile organisms which make their own food by photosynthesis and contain chlorophyll. Examples from Kingdom Plantae include trees, bushes, grasses, and algae.

Kingdom Animalia: organisms including mollusks, arachnids, insects, fish, reptiles, amphibians, birds, mammals, and humans. What these diverse creatures all have in common, besides being eukaryotes, is they are all capable of movement and they all ingest their food because they are not capable of synthesizing it.

The kingdom we are most interested in of course is the plant kingdom or Kingdom Plantae. This Kingdom is divided into two groups, the angiosperms and the gymnosperms. The gymnosperms contain nine further divisions called "Phyla." Plants such as conifer trees, ferns, mosses, gingko, cycads, and "living fossils" such as the Dawn Redwood (*Metasequuoia glyptostrboides*) and the Wollemi pine (*Wollemia nobilis*), are all gymnosperms which literally means "naked seeds." Most gymnosperms have needle like leaves, are evergreen, and their seeds form in cones. A few like the gingko and the bald cypress do drop their leaves but these are the exceptions rather than the norm.

The second group, the angiosperms, has only one phylum. Angiosperms generally have seeds enclosed in fruit, broad leaves, and a spreading habit. Most of the crop plants on which people depend for food are angiosperms. The single phylum, called Angiospermophyta is divided into two classes, the Monocotyledonae (or "monocots") and the Dicotyledonae (or "dicots").

Monocots have seeds with one seed leaf, leaves which are narrow with parallel veins, flower parts which usually occur in multiples of three, and no wood producing cells of which to speak. They tend to grow from the inside out, in the manner of these familiar monocots; grasses and palm trees.

Dicots have seeds with two seed leaves, broad leaves with central midribs and branching veins, flower parts in multiples of four or five, large colorful flowers, sometimes woody stems, and tend to grow in a branched habit, in the manner of these familiar dicots; oak trees and apple trees.

Each class is further divided into orders and then families. Sometimes it is easy to confuse these divisions, since both order and family names often end with the letters "ae" or "eae." There are many plants in which the order name and the family name are very similar, as in the case of the Welsh onion:

Kingdom: Plantae
Phylum: Angiospermophyta
Class: Monocotyledoneae
Order: Asparagales
Family: Alliumacae
Genus: Allium
Species: fistulosum

Each family is divided into genera, the singular of which is genus. In this way plants which are related can be described by the name of the genus, then by their species and finally by their variety, which is sometimes a natural variation and other times a "named cultivar."

Thus a taxonomic structure for classifying plants is born.

A date tree of the Deglet Noor variety for instance would be described as:

Kingdom: Plantae
Phylum: Angiospermophyla
Class: Monocotyledonae
Order: Arecales
Family: Arecaceae
Genus: Phoenix
Species: Dactylifera
Variety: Deglet Noor

So the binomial name of the date would be taken from the genus and the species names and this date plant is named *Phoenix dactylifera* with the name of the genus being capitalized and the species name written in all lower case letters and the usual custom is to italicize both the genus and species names. Plants described by three names are usually specific variations, or cultivars, the last name being the name of the variety, which may also be the name of a person or place or characteristic. For example, the Red Delicious apple's scientific name is: *Malus domestica* var. Red Delicious.

Inter-species hybrids may be indicated by an "x" or a multiplication sign between the two parent species. Thus the commercial strawberry, which is a cross between *Fragaria chiloesis* and *Fragaria virginiana* is sometimes described as *Fragaria* x *ananassa* or *Fragaria chiloensis* x *virginiana*. Rarer inter-generic hybrids may have a combined genus name with an "x" before it. An example of this is the grain crop triticale, an inter-generic hybrid of wheat (*Triticum* spp.) and rye (*Secale cereale*). Triticale is thus named × *Triticosecale*.

A graft-chimaera may be indicated with a plus sign, the name of the scion being written after the plus sign. An example of this is *Ziziphus spina-christi* + *mauritiana* for an Indian thorn plum scion grafted on a wild relative's rootstock.

There are several books which can help with scientific names and taxonomic structures and rules. One is the "Macropedia" volumes of the *Encyclopedia Britannica*. Another is *The International Code of Botanical Nomenclature*. *Hortus Third: A Concise Dictionary of Plants Cultivated in the United States and Canada* published in 1976 by the Macmillan publishing house, is also a valuable resource.

It is also useful to remember the family names of some of the more important species of trees, such as

Acer (the maple family), Plantanus (the Plane tree family), Pinus (the pine tree family), Malus (the apple tree family), Juglans (the walnut tree family) and so on.

It is important to know the walnut tree is related to the pecan tree, and the nectarine tree is related to the plum tree, the peach tree, the cherry tree, the apricot tree, and the almond tree. Not only does the knowledge of the relationship encourage the tree planter to try related species after successful plantings, but the relationships between species are necessary knowledge when top-working crop trees. In nursery propagation of trees a well marked sapling would have a colored tape or sign attached to it with the genus, species and varietal names of a fruit tree. Labeled in this manner the addition of the Latin name keeps mistakes from being made and time, effort and money wasted.

This system is by no means perfect, of course. There are still plants which have two or more scientific names as it has not been decided exactly where the plants belong in taxonomic terms. Sometimes the scientific name of the plant is changed and the change is slow in getting into textbooks and databases. This is why older plant designations are usually included when describing a plant in taxonomic terms. For example, the rare desert fig, *Ficus psuedosycamorus*, was recently renamed. The sub-species of this plant with lobed leaves has been designated *Ficus palmata* and the sub-species of this plant with the simple spade-shaped leaf has been named *Ficus cordata*.

Unusual plants and recent discoveries, such as the Wollemi pine (*Wollemi nobilis*), which qualifies in both categories being very unusual and discovered only a few years ago, sometime require the creation of a new genus. In the case of the *Wollemi nobilis,* this is because all the other species which might have been placed in this genus have been lost to time. All of the close relatives of this remarkable tree lived in the time when all the continents were connected to each other.

There are even some scientists who would like to add another Kingdom to the five which are currently the basis of the classification of all living things on earth. Many creatures live far below the surface on the ocean's bottom next to volcanic vents. These organisms do not directly or indirectly receive energy from the sun (unlike almost every other living organism). They get their energy from chemicals in the heavily mineralized and super heated water. Because of this dissimilarity, it has been proposed they be placed in their own Kingdom with other extremophiles. The giant tube worm for instance relies on bacteria inside its body to make organic molecules from oxygen, hydrogen sulfide, and hydrogen. This process is called chemosynthesis and is vastly different from the way most other living creatures receive their food.

This discussion is current and the creation of a new Kingdom of living things would truly be exciting. However, there is no guarantee such a change will occur. The entire subject is complicated by the fact these organisms are comprised of creatures from many different kingdoms, phylla, families and genera and include animals, plant-like and fungus-like organisms, and a vast amount of new micro-life.

There is also a movement to split the Kingdom Monera into two different kingdoms, the Eubacteria (true bacteria) and the Archeae bacteria, which could accommodate many of the extremophiles found in deep sea vents, hypersaline environments, frozen environments, and chemically harsh environments.

Does the ultimate source of their life energy make the newly discovered life forms so different they cannot be sorted out into the existing taxonomic kingdoms? There are convincing arguments on both sides of the question and the discussion will certainly continue as more extremophiles are discovered and researchers try to fit them into the classifications which may have to strain bit to contain their strangeness and diversity.

Thanks to Linnaeus though, at least we will be able to name them.

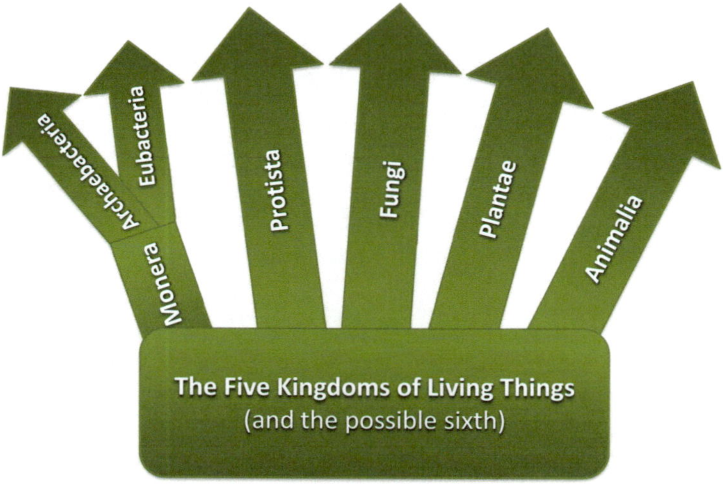

Five Kingdoms of LIving Things
Graphic by Sarit Rosenfeld/AIES

Chapter 12
Sugar Trees

Sugar crops such as cane, beets, and corn are water intensive crops, only suitable for cultivation in good deep soil and relentlessly destructive to the soil when grown as monocrops. Since these crops are not only being used to produce edible sugars but are now being converted into biofuels, the amount of hectares planted in these crops grows and grows while the competition between man and machine for the harvests becomes more acrimonious and controversial. So far in the competition for sugar crops between humans and the automobiles, the automobiles seem to be winning.

This is ridiculous, of course, as diverting all the grain and sugar crops grown in the world could not fill the energy needs of the millions of internal combustion engines—and if it did who would be left to drive them? It is estimated the sugar and grain crops used to produce a tank of ethanol would feed an adult human being for 14 months. It may be some time before the cruelty and futility of this policy becomes apparent and sugar and grain crops are crossed off the lists of sources of ethanol and other biofuels. Meanwhile permaculturalists and orcharders should promote the planting and cultivating of "sugar trees," which are much less demanding as to water and soil and can be grown on marginal and depleted land and actually help regenerate these areas while producing delicious and usable sugars.

Maple Trees (Acer spp.)

Certainly one of the most delicious products of the sugar trees are the syrups and sugar produced by the maple tree. Of the one hundred species of maple in the world most will yield a good tasting sugar or syrup. The exceptions are the ornamental maple trees whose milky sap is not suitable for syrup or sugar production. Sugar and syrup are most commonly made from the maples which have the highest percentage of sugar in their sap. These are the sugar maple (*Acer saccharum*), the red maple (*Acer rubrum*), the silver maple (*Acer saccharinum*) and the Manitoba maple or boxelder (*Acer negundo*).

Maples are easy to identify because of their symmetrical branching habit, the shape of their leaves, and their unique winged seeds called "samaras." Maples are large deciduous trees which grow in moist deep soil in temperate areas of the world. The sugar content of the sap varies a good deal from tree to tree, type to type, and area to area. Sugar content of the sap can also vary by the time of day.

In the later days of the summer and fall the maple trees stop growing and begin to store excess starches in specialized cells called ray cells. This excess starch remains stored during the cool months of the winter. When the weather warms above 40° F the starches change to sugar, mostly sucrose. These sugars pass into the sap. At temperatures at or above 45° F, sugars are no longer produced. So the season to collect maple sap is short, the sugary sap only flows in early spring when the temperatures are between 40° F and 45° F.

The rising temperatures also create pressure inside the tree. When holes are bored into the tree the sap drips out. Trees are "tapped" in this manner when they reach about 10 inches in diameter. At that point a tree can bear one tap. Trees 20 inches in diameter can sustain two taps and trees over 25 inches in diameter can sustain three. No tree should have more than three taps or it would be damaged by having too much of its sap extracted. Trees with large spreading crowns are usually the best producers.

Fresh tap holes are made each spring and the tapper takes care not to drill near an old tap hole and not to bore more than 1 and a half inches deep. The tap holes are made with a sharp bore bit to minimize damage to the wood and done at a slightly upward angle so the sap can flow freely. The bore bit should be the size of the spouts chosen for the tapping project. In areas where manufactured taps are not available tree tappers often hand-carve wooden spouts.

The spouts are tapped into the bored hole so tightly it is not possible to extract them by hand. Taps are inserted on slightly warmer days when there is no danger of splitting the tree. A bucket or container is then hung under the spout. The bucket should have a cover to keep out rain, snow, and other unwanted material.

A taphole in a healthy tree may yield 5 to 20 gallons of sap. Ten gallons of sap must be boiled down and reduced to yield one quart of maple syrup. This is usually done outside or in well ventilated areas because the sap, as it is reduced, lets off a lot of steam. The sap should be collected every day and kept cold before it is boiled as it can sour in the heat. The syrup is finished when it reaches 66-67% sugar. The syrup should be canned hot. Open containers of syrup must be refrigerated. If the reduction process continues the sap is reduced to the point where it makes granules when cool. At this point it can be put in molds to make solid bars of maple sugar.

This method of taking sap has been used for many years and means hundreds of buckets have to be collected and brought to a central point. This has always been hard work in challenging conditions. In the 1950's the first experiments in collecting sap by vacuum tubing were set up and proved to be effective. One taphole was made per tree and a plastic tube was attached which led to a central collection point. In the 1960's low vacuum was applied (up to 15" Hg) through the tube. This level of suction allowed sap production to double with no harm to the tree and has been used for over four decades in the US and Canada. There are still many areas, however, where sap collection is accomplished in the described traditional manner.

The maple trees themselves have great value as timber producers, elements of erosion control, land reclamation in rocky areas, soil enrichment for cool areas, and for their stunning autumn coloring and their symmetry. They are among the most beautiful trees in the landscape.

Birch Trees (Betula spp.)

Birch trees are deciduous, extremely cold tolerant, flexible, and wind-hardy trees found in the northern latitudes and temperate zones. Birch trees yield wonderful syrup with a distinctive caramel-like spiciness. It is consumed much the way maple syrup is, in sauces, in glazes, on pancakes, and it used to flavor soft drinks, beer, and ice cream.

Birch sap is fundamentally different than maple sap in its composition, being mostly fructose and glucose with a small amount of sucrose and galactose. Birch syrup is harder to make than maple syrup as there is less sugar by percentage in birch sap. Eighty liters of sap may become one liter of syrup. Reverse osmosis techniques are sometimes used to remove the water in the sap and thereby concentrate the sugars as it is easy to scorch the delicate and varied sugars in the birch sap during the reduction process.

The opportunity for tapping birch trees is even shorter than the one for maple trees because birches live primarily in colder areas than maple trees. The birch trees are tapped a few weeks before the leaves appear commonly in the beginning of April. Birch trees are tapped closer to the ground than maple trees. Usually only one or two taps are placed, even in the largest trees, as birches have a lower root and trunk pressure than maples. Most commercial birch syrup is produced from Paper Birch tree (*Betula papyrifera*), Alaskan Birch trees (*Betula neoalaskana*), and Kenai Birch trees (*Betula kenaica*). Other species of Birch are tapped in Scandinavia and Russia.

Birch trees are also valued for their timber, withies, their useful and flexible bark, and the value as a land reclamation tree in cold and windswept areas.

Hickory Trees (Carya spp.)

The shagbark hickory tree (*Carya ovata*) can be tapped as well, though the yield of syrup is usually low and the syrup collecting window is rather short. A more popular way of making hickory syrup is boiling bark or nutshells from the shagbark hickory into a smoky tasting liquor, then sweetening with sugar and reducing it to syrup consistency. This unusual local syrup is also whipped up into a spreadable "butter" and is surely one of the most unusual products of the sugar trees. Oddly enough, the leaves and twigs of the hickory trees have been boiled and the resulting liquid dried to make a substitute for table salt. So it might be said the hickory tree is both a salt and a sugar tree.

Poplar Trees (***Populus spp.***)

Poplar bark, once a flavoring for spruce beer and sometimes used as a stimulant for the heart, is used to prepare poplar syrup, in a manner very similar to the use of hickory bark. This essentially means poplar bark is the flavoring for a type of sugar syrup rather than a source of sugar. Still the syrup has become popular as a gourmet item and some experimentation has been done without notable success with taping poplars for sap. There does seem to be usable sap in some kinds of poplar but currently there does not seem to be an efficient way to collect it.

Other Nut Trees

Walnut trees (*Juglans spp.*), black walnut trees (*Juglans nigra*), and pecan trees (*Carya illinoinensis*) can also be tapped in areas with the proper seasonality, allowing a sugar collecting period in the spring. Yields are also low with these trees but the resulting syrups are very flavorful. Black walnut syrup is considered to be superior in taste even to maple syrup which is more popular and available by far. The low yield and delicate flavor of these syrups have led to many experiments in Reverse Osmosis treatment of the sap which removes water by forcing it though permeable membranes, leaving the syrup behind and thickening it without heating it.

Carob Trees (*Certonia siliqua*)

The carob has been food for humans and animals around the Mediterranean Sea area since prehistoric times. It is an evergreen tree with glossy leaves. Carobs are tolerant of aridity and salinity but sensitive to frost. Sometimes the carob tree is the only tree living in steep, dry, and arid areas and so, the only source of food. Luckily the pods of the carob are highly nutritious and with other vegetation can provide the basis for feeding sheep, goats, horses, and even cattle. Carob trees copiously bear brown, woody pods with small seeds of amazingly regular size and shape. Even trees grown in extreme conditions regularly produce fifty kilos of pods. Trees in well watered, richer soils produce literally hundreds of kilos of pods.

As can be imagined, since carob trees are leguminous, and also so very resistant to challenging conditions and so very good for the soil, they have now been planted in poor and marginal areas all over the world. In addition to the carob's value as a reclaimative, anti-erosive, and fodder providing tree, the pods can be made into delicious food for human beings. Bread made with 25 percent carob flour, carob cake, candy, and syrup have all made their appearance in health food shops. It is actually the tissue between the hard seeds which is ground to powder and used a chocolate substitute or milled into flour. Baked goods made with carob flour are rich and nutty tasting. The yield of carob syrup is very impressive. After slow cooking of the whole pods, the yield of syrup is almost a kilo of syrup per kilo and a half of pods. In Turkey this is the most popular local sweetener after native honey.

In folklore the carob is a symbol of longevity but the carob's reputation for a long juvenility period, forty years before the carob bears fruit, is totally undeserved. Carob trees, given a reasonable amount of water and care, flower and bear fruit in the fifth year, like most other fruit trees.

Honey Locust (*Gleditsia triacanthos*)

This is another leguminous tree from southeastern United States which is both useful as a stock feed and has great untapped potential as a sugar tree. It does not do well in truly arid areas but can be grown in marginal soils, poor soils, rocky soils, and soil containing clay. The tree grows steadily under most conditions. It is tolerant of low temperatures, dampness, and short periods of drought. It is also tolerant of alkalinity. While a honey locust tree does not exhibit the root nodes of another promising tree in the same family, the black locust (*Robinia pseudoacacia*), the honey locust is associated with both nitrogen fixing soil symbionts and beneficial fungi.

Because of the tree's extreme hardiness the honey locust tree was chosen in 1934 for Franklin D. Roosevelt's great Shelter Belt project, a very large tree planting project carried out on 33,000 farms to protect the soil and mitigate the heat and drought conditions of the "Dust Bowl" years. Of the trees planted, which included black locusts, Siberian elms (*Ulmus pumila*), and cottonwoods and poplars (*Populus spp.*), the honey locust trees had the highest rate of survival.

As a fodder tree the honey locust is very productive, bearing on average a hundred kilograms of pods per tree which are very attractive to most farm animals, especially pigs and cows. Chickens can be fed with ground or crushed pods. Sheep, goats, and horses will reach over fences and push through hedges to get to the fallen pods.

In the 18th, 19th, and early 20th centuries, North Carolina honey locust pods were made into meal and added to baked goods instead of sweetening. Ground pods of this tree make an excellent flour extender and can be added to bread, cakes, and muffins. Children often eat the sweet pulp in the pods. The yield of syrup is not as high as the carob but at least as high as the mesquite with less of an astringent edge and more of a honey like flavor.

The black locust tree exhibits some of the qualities of the honey locust. It is a hardy tree, with excellent hard wood, a nitrogen fixer which exhibits the nodes of a legume on its extensive root system, a valuable shade tree, and a possible sugar source as some black locust pods contain up to 30% sugar. It has one cautionary point however; the pulp in its pods contains a variable amount of toxins making them generally inedible. This makes it, at present, not a reliable source of sugar nor a reliable fodder as there is too much of a chance of utilizing the products of a tree with high toxicity. Breeding and selection could solve this problem.

Surely the potential of both these locust trees should be investigated and these trees integrated into perennial agricultural formats as soon as possible.

Mesquite Trees (*Prosopis spp.*)

The hardy mesquite tree, famous for the density and quality of its wood and its nitrogen fixing and soil improving qualities is also a "sugar tree." Mesquite trees produce pods full of protein and sugar. These pods are crushed and boiled to yield syrup or ground to yield sweet gluten-free flour. The three most common mesquite species yield useful pods, the honey mesquite (*Prosopis glandulosa*), the Screwbean mesquite (*Prosopis pubescens*) and the Velvet mesquite (*Prosopis velutina*). Useful pods can also be obtained from the Common mesquite (*Prosopis juliflora*) as well. There is some discussion about the suitability of pods from the Black mesquite (*Prosopis Nigra*) for making syrup or meal, some of which are too bitter to be palatable as food for humans. Black mesquite pods may not be the best fodder for animals either, as sore mouths and other problems are sometimes seen in areas where the black mesquite is the primary fodder tree.

Mesquite meal is made by sorting the pods and discarding those which are light weight or broken. The full undamaged pods are then heated slowly until they become brown and crumbly. This can be accomplished by drying in full sunlight or placing in a low oven. Then they are ground into meal using a traditional metate or a modern food processor.

Mesquite molasses is made by adding one pound of mesquite pods, washed and sorted, to 4 quarts of water. The water and pods should be cooked in a covered pot at low heat for 12 hours, strained, and then the liquid is reduced to syrup.

Palm Tree Sugars

Some of the very tastiest sugars come from the sweet sap of palm trees

Quantities of sap which can be collected vary from tree to tee and species to species. An ordinary palm may give 5 kilos of sugary sap per day from a few inflorescences. A high yielding tree may produce 20 to 25 kilograms from multiple flower columns. Some trees have been tapped for decades without harm as long as they are well watered. The sap can be collected for several months from both male and female trees with generally a longer tap flow from the female trees.

Palm sugars are usually harvested from local trees and so have local names. In Burma for instance, sugar from the coconut palm is called “jaggery.” In Indonesia the same kind of sugar collected from the same species of trees is called “gula merah.” Since the composition of the sugars vary according to the species from which the sap is collected (and sometimes by the time of year within species), palm sugars are generally composed of sucrose, fructose, and glucose in varying percentages and compositions.

The importance of the sugar palms in small Asian farms is often underestimated. In mixed farming operations the trees provide sugar, fruits, germinated seeds, juice for human consumption, and animal feed. They make up the green and living fences around households and grow on the dikes around flooded fields. They also provide shelter for birds and bats, which are essential for the health of the farm. Their leaves are useful as thatch or twine and waste from the tree is burned as fuel.

Palm sugar is widely used in Asian cooking to sweeten savory food, sweeten baked goods or to balance out strong or salty flavors.

"Toddy"—or "treetap"—is the general name for the mildly fermented or fresh sap of a wide variety of Asian palms. The sap is collected in the early morning by a tapper who climbs up to the crown of the palm tree where he has set one or more taps in cleaned areas on the trunk of the tree. The fresh sap is poured from the tap's attached gourds or jars into a covered, larger vessel, usually worn on the belt of the tapper or carried with a shoulder strap. It will be served by the afternoon in a local toddy house. Most treetaps begin to sour in the heat and dampness after a few hours or are turned to alcohol by indigenous yeasts. Toddy is rarely turned to syrup or sugar but utilized more as a local "soft drink" in many countries. Some treetaps are allowed to ferment and are then distilled into a truly foul tasting alcoholic drink, much like rank tequila. But on the whole, treetaps are usually consumed within a day of tapping.

Tapping may continue from individual trees for many months, but taps are usually removed from the trees before stormy and cool seasons. They are replaced in new holes when the trees are invigorated by rain and warmer weather. Other species are tapped by cutting into emerging flowers. When tapping is finished the flowers may be cut off at the base so new flowers will grow from the leaf junction, node, or the crown of the tree.

A familiar tree which is tapped by cutting the inflorescences is the coconut, probably the most widespread of the palms in the tropical world and valued mostly for its large seeds and sweet "milk." The fact the seeds are so highly valued precludes tapping for sugar in most places, but in areas where there are coconuts to spare, the sweet sap may be collected from tapped flowers, usually for immediate use, rarely fermented or concentrated.

"Sugar palm" is the common name for several species of palm which are tapped for their sweet sap. These include the Arenga palm (*Arenga pinnata*), Palmyra palm (*Borassus flabellifer*), and the Toddy palm (*Caryota urens*).

The Nipa Palm (*Nypa fruticans*) and Coconut Palm (*Cocos nucifera*) both produce sugary sap. The sap from these trees is collected by cutting the flower buds or slitting the underside of the flowering column and then collecting the sugary sap. The sap is boiled until it thickens. Then it is put into bamboo tubes or simple molds to harden. Since palm sugar comes from a variety of sources and is processed by cottage industry there is a great deal of variability in the color, taste and composition of the sugar.

The Arenga Palm (*Arega pinnata*) is a source of a sugar known as gur. It is a medium sized palm growing to 20 meters tall with a rough trunk covered with the stubs of old leaves. The leaves are very large, 6 to 12 meters long and 1.5 to 2 meters broad. The fruit of the tree is about 7 cm in diameter, sub-globose, green but darkens as it matures. The sap is harvested for commercial use, made into sugar, and also fermented into vinegar and wine. The immature fruits are eaten in Southeast Asia and the Philippine Islands and sometimes canned after boiling in syrup. Both raw juice and pulp of the fruits are caustic in nature, while the sap is very sweet and sometimes drunk as fresh toddy.

The Palmyra Palm (*Borassus flabellifer*) can reach a height of 30 meters and live 100 years. The trunk of the tree resembles the coconut palm's trunk, smoother than the trunks of most palm species but ringed with leaf scars. The young inflorescences, both male and female, yield sugary sap which is fermented into an alcoholic beverage called "palm arrack" or concentrated into a crude sugar called "Gula Jawa."

The fruit, borne in clusters, has a black husk and measures four to seven inches in diameter. The top portion of the fruit is cut off to reveal three sweet translucent jelly-like seeds, similar to lychees, but having no pit. The ripe, fibrous, outer layer of the palm fruit can be roasted, boiled, or eaten raw. Fresh sap is considered a laxative. The underground stems of sprouted seeds are eaten much like hearts of palm, as are the cooked sprouts of young plants. The huge leaves are used for making baskets, mats, fans, hats, umbrellas, and as writing and thatching material.

Unlike most palm species, the timber from old Palmyra palm trees is hard, heavy, black, and durable, much valued for construction. This is truly one of the most beautiful and useful palm varieties.

The Toddy Palm (*Caryota urens*) is sometimes called the "Wine Palm." It is a rain forest palm from Sri Lanka, Myanmar, and India with a single trunk and rather small in relation to other palm species, usually about 12 meters in height. It is also relatively short lived, dying after flowering and fruiting. To collect toddy for wine and sugar making, and also to prolong the life of the palm, the numerous inflorescences which emerge at each leaf node are tapped and the sap is collected. The total amount of sap flow from so many flowers may also lead to the death of the tree, but fruit formation, with the production of the round, 1 cm in diameter, red drupe, which is saturated with oxalic acid, is sure death for the palm. The palm trunks have no value as timber but the leaves are used to make wine baskets and thatch.

Date Palms (Phoenix dactylifera)

These large stately palms are found all over the Middle East and Southern Asia. Date palms are also actually among the palms tapped for sugar. A vertical slit is made in the base of the inflorescence and a small piece of wire or wood is inserted to keep the slit from closing. A receptacle for the sap is hung underneath and several liters of sap can be collected in this manner. The sap is mostly sucrose, clear and only slightly sticky. This sap turns milky and begins to ferment within a few hours. A more drastic form of tapping is done by clearing away many upper leaves and tapping the date trees trunk. This is not a common practice with the domesticated date tree since careless tapping can kill a very productive fruit tree and even careful tapping of the sap reduces the yield of a tree for 3 to 4 years until the tree recovers. The wild date tree is often tapped for its sap using this method as the fruit is often of little value as food. Hundreds of tons of sugar are obtained in this manner from wild date trees in India where expert tappers extract sap from male and female trees for periods of up to 60 days while doing little harm to the trees.

Most date sugar and all date syrup are made from the fruits themselves. Thick brown syrup called “silwan” is made from soft desert dates. It has a honey like consistency and a deep rich flavor.

Dates are different from many other fruits as their sugars are too concentrated to be pressed out. This means water has to be added during the extraction process. So silwan is made in two different manners. Both involve the use of added water.

The syrup can be made in the traditional manner, by boiling dates in water and extracting the juice from the boiled dates, an action which is repeated up to four times until most of the sugar has been removed. The presscake, consisting mostly of fiber and seeds is often fed to animals. The strained juice is then reduced until it reaches a desirable consistency.

The modern industrial process involves crushing the dates and mixing them with water. The mixture is then heated, cooled, and pressed. The resulting juice is strained and reduced. More syrup is obtained in this manner and it is of better quality, though there are many people who prefer the traditional product because of its more robust taste.

"Golden sugar" is made from overripe and damaged dates which are pressed for their juice after boiling and cooking. The resulting liquid is strained many times and the sugar obtained by reduction. Golden sugar is moist and crystalline, as is brown sugar, but it has a distinct date-like taste. Date sugars invert easily with more and more sucrose converting as harvest time approaches.

There are other sugar trees, of course, many known only locally. The desire for sweet foods seems to be a basic human desire. This desire can be satisfied—and many other benefits gained—simply by planting these marvelous sugar-source trees on the steep, rocky, marginal, boggy, arid, worked-out, and abandoned areas of the world.

Chapter 13
Salad Trees, Tree Vegetables, and Leaf Protein

Among the woody plants and trees of the world which are used for food, there are some with edible plant parts and leaves which can be eaten as salads or vegetables.

They are generally underutilized and sometimes not even recognized as possible food sources. Some of them are edible from crown to root such as the Moringa tree which is in a class all by itself.

Moringa oleifera is the most studied of fourteen types of Moringa trees. They are sometimes called the "Never-Die Trees" and are also called the "Drumstick Trees." The *Moringa oleifera* is a small, fast-growing tree which can live for up to 20 years. Its growth rate is truly amazing as it can reach 3 meters in height in its first year of life. The tree has deep roots, allowing it to survive in dry regions. The wood of the tree is soft, as are its small round leaves, which are tender enough to be eaten straight from the tree.

From the *Moringa oleifera* comes edible leaves, edible flowers, pods which can be stir-fried when they are green (the "drumsticks"), growing tips which somewhat resemble celery stalks, seeds pressed for oil when the pods are mature, plus roots for pickling.

The tree is called "Nebeday" in Senegal, most likely a derivative of the English words "never die" in reflection of the tree's ability to withstand drought, grow quickly from seeds or cuttings, and regenerate itself even after the most severe pruning or coppicing. It is sometimes cultivated as a living fence around gardens. Although the tree is esteemed for its many medicinal uses, including control of blood sugar and reducing blood pressure, it is not commonly known the leaves of this tree are extremely nutritious, and the pods, flowers, and growing tips of the tree are also edible.

The traditional method of preparing Moringa leaf sauce causes much of the leaves' nutritional value to be lost. Fresh leaves are boiled two or even three times, with the water discarded each time, in order to remove the leaves' somewhat bitter spicy taste. The raw leaves however, have large amounts of Vitamin A and C and contain calcium, iron, and other minerals which may be lost by boiling and leaching. A better way to preserve the nutritive value of the Moringa is simply to dry the leaves and grind them into powder. This powder may then be turned into sauce or added to soups to thicken them or to add taste to staple foods.

The benefits of using the Moringa in this matter are many. Fifty grams of dried Moringa leaf powder can supply 80% of the protein needed in a child's diet per day, as well as 100% of the calcium, 100% of the magnesium, 80% of the potassium, 100% of the iron, 100% of the vitamin A and 50% of the Vitamin C. Because of this and moringa's richness in amino acids, the Moringa is sometimes referred to as a "superfood." Many NGO's are working to get Moringa into the food systems of famine haunted nations.

Adansonia digitata is the scientific name for the huge Baobab Tree, also called the "Lemonade Tree," the "Cream of Tartar Tree" and the "Monkey Bread Tree." The trunk of the baobab often reaches an enormous girth. Its odd, root-like branches and spreading crown have given rise to the African myth the tree was planted upside down by God.

The baobab thrives in desolate and drought-prone plains, sometimes the only tree in the landscape and often the keystone species of the ecology of arid and semi-arid areas. There are five species of baobab, four in African and one in Australia. The most useful one from the point of being accessible and edible is the *Adansonia digitata* or Common Baobab.

The Baobab is a deciduous tree. Its large pendulous flowers, which are pollinated by bats, support the lives of bats, birds, and many types of insects. Animals of all kinds benefit from dropped flowers, fresh leaves, and baobab fruit. The fruit when ripe is a large woody capsule, roundish and covered with short yellowish brown fuzz. Inside, the fruit has a dry powdery layer which covers hard black seeds inside a woody shell. The powdery substance is rich in Vitamin C and potassium tartrate, and is sometimes soaked in water by local people to make a refreshing drink. This substance has given the trees its nicknames of the "Cream of Tartar Tree" and the "Lemonade Tree."

The Baobab contributes in many ways to the food systems of local people. Fresh leaves are used as ingredients of soups. Edible caterpillars, an important source of protein in Africa, are fed on the leaves. Also, the hand-sized leaves and fallen flowers are gathered to feed cattle. Dried leaves are added to sauces, stews and gravies. The immature pods, which somewhat resemble green lemons, are eaten as vegetables either cooked or raw and are also dried in slices to be added to cooking foods later. The young sprouts of baobab are eaten like asparagus as are the roots of young trees. Both sprouts and roots can be pickled. The hard black seeds are ground as a coffee substitute.

Besides supplying the flavoring for baobab "Lemonade" baobab trees sometimes supply the water and the sweetening as well. Rain water collects in the hollows made by the clefts of large branches. This valuable source of water is often used by local people. Hollows in the trunks of trees are utilized by African honey bees and some trees have been fitted with pegs or ladders so honey can be gathered in season. In particularly dry years the tips of branches or slips of baobab wood are also pressed or sucked for moisture.

Powdered bark and powder from the mature Baobab fruit have reputed medicinal properties and are used to treat skin conditions and reduce fevers.

While there is apparently great variability in the nutritive value of leaves and fruit from the baobab tree, the minerals and vitamins present in all the baobabs' plant parts: iron, potassium, calcium, and Vitamin C, make valuable additions to local diets.

***Leucaena* spp.** are fast growing leguminous trees from South America. The Leucaena tree was promoted as a miracle tree and planted in great numbers to produce fodder, biomass for biodiesel production, and also to control erosion and to reforest desolate areas. It grows especially well in warm climates. In some places it became invasive and crowded out native vegetation but most of the plantings were successful and the tree proved its value.

While it is considered a multipurpose tree with an emphasis on its importance in animal feeding and biofuel production, this is another tree with edible leaves and pods. Leucaena foliage is excellent fodder for animals but the young leaves and tips are tender and digestible for people to eat. The long green pods can be boiled or eaten raw.

While the mature pods are too tough and papery to use as food, the seeds of the *Leucaena esculenta* are edible and can be roasted and eaten. Most other leucaena species have pods which are mildly toxic to non-ruminants and should not be eaten. Leucaena leaf powder and dried young Leucaena leaves can be used in soups, stews and sauces very much as moringa leaves are used in Africa.

Since leucaena plantings were set out in many very poor countries, its immediate usefulness to hungry people should not be forgotten.

A short list of common trees with edible leaves

Most people can name ten common vegetables with edible leaves, such as amaranth, beets, collards, spinach, mangolds (beet greens), and arugala. Few people know many common trees have edible leaves. Below is a short list of common trees with leaves which can serve as food for the hungry. Most are plants with other uses and stripping or gathering their leaves for food is only done in times of famine. In other cases it is simply not generally known the leaves are edible, so the local population does not eat them.

Acacia spp. (selected species): The three acacia species which have edible leaves are *Acacia coccina* (the Soap Pod tree), *Acacia famesiana* (Sweet acacia) and *Acacia nilotica* (Prickly acacia). None of these leave are particularly appetizing. Young leaves taste better than mature leaves and the best of the lot is the "Sweet" acacia. Edible leaves have to be picked out between the long thorns of *A. nilotica.*

Albizia julibrissin (the silk tree): This tree has rather feathery mild tasting leaves which are easy to chew. The pods are bitter and not attractive.

Aralia elata (the Japanese angelica tree): This is a large, deciduous, and frost tender tree. It is valued in Japan for its beauty but also because the young leaves and shoots are edible. The first leaves and shoots of the spring are blanched and eaten like asparagus. More mature foliage is steamed or stewed.

Berberis spp. (selected species): Several members of the Barberry genus have edible leaves and some have edible fruits. *Berberis lyceum* has rather salty edible leaves. *Berberis vulgaris* leaves are astringent and chewy. Barberry plants produce berries with medicinal value which are eaten by birds and other wildlife. They can also be used to make conserves and jellies, dried, or eaten raw after a frost, which reduces their acidity.

Cordia spp. (selected species): There are several *Cordia* species with edible leaves, including *Cordia myxa* and *Cordia dichotoma*. It is best to carefully and accurately identify local species before sampling them. Inedible *Cordia* species have leaves which are extremely bitter and may contain anti nutritional elements such as tannins.

Fagus spp. (selected species): *Fagus* is the genus name of the Beech trees. *Fagus grandifolia* is the American beech tree. The name of the European beech is *Fagus sylvatica*. The nuts of the beech tree are small, easy to crack, and sweet. The new young leaves are edible and tasty, but the older mature leaves are much harder to chew and leave one with a dry mouth.

Morus spp. (selected species): Morus is the genus of the Mulberry trees. Mulberry leaves are favorites with animals and insects. They are the food of the most famous caterpillar, the silkworm. The very young leaves are juicy and mild tasting to the hungry human and a favorite food for many leaf eating creatures. In order to get a taste of an untouched mulberry leaf, it is best to pick the newest leaves early in the morning and be sure to get there before the competition. Warning: do not eat green mulberry fruit as it can be poisonous.

Pistacia spp. (selected species): Only eat very young leaves of *Pistacia terebinthus* and *Pistacia chinensis* trees.
Prunus spp. (selected species): Only eat very young leaves of edible-leaved prunus species, *Prunus padus* and *Prunus tomentosa.*

Ribes spp. (selected species of "currants"): The leaves of these species of *Ribes* are edible: *Ribes cereum*, *Ribes divaricatum*, *Ribes odoratum*, and *Ribes nigrum.* Are all chewable and mild tasting, but they should be scalded and rinsed well before eating as they are also the favorites of local insects.

Sambucus nigra: The leaves, flowers and berries of the elderberry, *Sambucus nigra* are all edible. The very young leaves can be eaten raw. The flowers are delicious fried in batter. The berries can be eaten fresh, turned into jellies and conserves, or made into syrup and wine.

Spondias spp. (selected species): The leaves of the yellow mombin (*Spondias mombin*), the Otaheite apple (*Spondias dulcis*) and the jocote (*Spondias purpera*) are edible. However, they are not very tasty raw and I have not found any recipes for cooking them.

Tilia spp. (selected species): The leaves of the Linden tree are large, thin, and bright light green in color. They can be eaten raw when young. They have also been used to wrap other foods such as cheese and fruit. They can also be wrapped around meat or grain mixtures and steamed somewhat like corn husks in the preparation of tamales. However, the cooked linden leaves are tender and edible unlike the corn husks. The best edible species are *Tilia americana*, *Tilia japonica*, *Tilia* x *europaea.*

Toona sinensis: *Toona sinensis* is a large tree native to Asia with edible fruit, shoots, and leaves. It is a popular ornamental tree, sometimes called the Chinese ailanthus. The fruits are small and sweet, usually red or reddish bronze in color. The young shoots can be steamed and taste much like spring onions. The leaves can be eaten raw or cooked and are high in Vitamin A.

Vitis spp. (selected species): Grape leaves are most often used as a wrapping for meat, rice or fish. As the food inside them is steamed or baked the leaves becomes soft and edible. They can also be shredded and added to soup. Picked when young and tender, they are sometimes preserved in oil until needed. They are also used to make wrappings for cheeses or confections. Dolmas or stuffed grape leaves are the most common way of using these leaves.

Ziziphus mauritiana: *Ziziphus mauritiana* leaves are dark green on the upper sides and light silver green on the undersides. The leaves are at their best when both sides are light green, usually in the early spring, before the characteristic fuzz forms on the leaves' undersides. The fruit of this tree is excellent, tasting like small sweet apples when ripe and very much like black cherries when dried.

Leaf Protein

There are comparatively few species of trees with leaves people can digest compared to the vast number of species available as animal food. However, there are many leafy plants which when pressed and processed can yield useful food for humanity in the form of leaf protein. Leaf protein is potentially an abundant source of protein for humanity as there are so many sources from which to extract the protein from the leaves. However, there are serious obstacles in both the production and utilization of this food source. The human digestive system cannot deal with the sheer bulk necessary for the digestion of most leaves and fodder crops, which contain much fiber and relatively little protein.

In order to overcome this physiological fact, leafy crops such as alfalfa can be pulped and their juice extracted. This leaves most of the bulk and fiber behind and the pulp can then be used for animal feeding. The juice is then heated to coagulate the protein, which is strained out and the resulting residue powdered. Leaf protein concentrate (LPC) is rich in amino acids, polyphenols, and vitamins, and should be a potential "superfood."

While experiments in leaf protein production have been ongoing for the last fifty years, the uses of LPC (leaf protein concentrate) have been few and very little has been used as food for the hungry people of the world. There are some very good reasons for this. As currently extracted, LPC has several anti-nutritional factors, including high percentages of tannins, phytates, and cyanide. Removing the anti-nutritional factors would make leaf protein more expensive by the kilo than steak. So despite early promise, the promotion a Nobel Prize winner, and much research, LPC remains only a potential food source for people, rather than an actual one.

The early excitement for LPC is reminiscent of the perceived promise of spirulina, the blue-green algae which would supposedly feed the world. The use of spirulina is limited by the human body's tolerance for nucleic acid. In the same manner, the use of LPC is limited by its anti-nutritional factors.

So far LPC has proved most valuable as a supplement in animal feeding. LPC is seen as a possible substitute in animal nutrition for crops humans might consume. However, since most animals can digest leaves, grass, and agricultural waste without the extraction process, the economic reasons for doing this are at best unclear. Chickens, for instance, gain weight more quickly and are healthier, if they are fed freshly cut alfalfa, than if they are fed LPC from alfalfa plants as a supplement. Undernourished children in Africa gain weight and regain their health much faster if they are given powdered Moringa leaves or whole fresh leaves as a food supplement than if they are given extracted leaf protein from the tree.

Still the idea remains an interesting one. If edible leaf protein could be extracted from leafy crops, tree leaves, and the by-products of harvesting, we could realize the vegetarian dream and literally live on grass. A fascinating book written by Nobel prize winner N.W. Pirie called *Leaf Protein Concentrate and its By Products in Human and Animal Feeding*, Cambridge University Press; 2 edition (February 27, 1987), outlines the problems and describes the great potential and possible uses of LPC.

Chapter 14
Trees That Changed the World

The Fever Tree (*Cinchona spp.*) And Malaria

This tree changed the world by saving millions of lives. Cinchona is also called Cinchona Bark, Jesuit's Bark, Peruvian Bark, Fever Tree Bark, and Quinine. The Fever Tree changed the world by reducing and in some cases eliminating the threat of one of humanity's worst enemies: malaria. Malaria is a disease which had killed hundreds of thousands, from the Arctic Circle to the water meadows of Italy as well as in tropical regions. It was a real danger for most of the world's population, taking a terrible toll on explorers, colonists, sailors and soldiers as well as indigenous populations.

The Cinchona is a tall striking evergreen tree which often reaches 30 meters in height when mature. The leaves of the cinchona are flat and broad, marked off by large veins and having a shiny dark green surface. Cinchona flowers are white, pink, or red in color. They are also thickly covered all over with silky hairs and emit a sharp odor when crushed.

The name of the tree, "Cinchona," reportedly comes from the name of a Spanish Countess, the Countess of Chinchon, who was treated for a fever using a remedy made from the bark of this tree while in South America. When she recovered, the countess sent saplings of the plant to Europe for investigation and planting in gardens of noble friends. She was convinced of the plants curative powers and convinced others to plant the tree. Because of her enthusiasm, the plant was named after her, or so the story goes. Unfortunately, this story is more fable than fact. The Countess Chinchon was the wife of the

Viceroy of Peru. She died suddenly on the way back from Peru to Madrid and did not seem to have had any illness resembling malaria. The Viceroy himself seems to have had malaria several times but there is no mention of a bark or a cure in his secretary's detailed diaries. He even had bouts of the disease when he came back to Spain. Apparently the curative powers of Cinchona bark were initially revealed to Europeans by Jesuit priests who traveled to Peru in the 17th century and learned about Cinchona from the native peoples.

Cinchona trees thrive in rain forests at elevations between 3,000 and 11,000 feet in South America. There are at least forty documented Cinchona species in South American forests. However, not all Cinchona types are potent enough to cure malaria, though many of the lesser species are used for other purposes.

The Jesuits found the native people chewed on the bark of the Cinchona to prevent shaking and chills from working in the harsh wet conditions of the colonial mines. The native peoples had traditionally used the bark for treating ordinary fevers and chills, bone aches, and sweating, as well as malaria. The first mention of the tree in literature as a cure for "ague" was written by an Augustinian friar and herbalist, Antonio de la Calancha, in 1638. Another priest, the Jesuit Bernabé Cobo, wrote a chapter in his work of many volumes, *The History of the New World* in which he describes the tree. The Jesuits first proved, then promoted the remedy and eventually Cinchona Bark, renamed Jesuit's Bark, became a very valuable commodity, which was literally worth its weight in gold.

Malaria was endemic in Europe and around the globe, killing and debilitating hundreds of thousands. It was thought to be caused by "bad air" hence the name malaria. Malaria is actually caused by the protozoans belonging to the genus *Plasmodium*, which are injected into the human blood by way of the sharp proboscis of a certain type of mosquito. The parasites attack the red blood cells once in the bloodstream bringing on the fever and chills of this debilitating disease. Cinchona bark, which acted very quickly, was therefore one of the earliest "miracle cures" obtained from the wild. However, the majority of European physicians still refused to use the new medicine on their patients as the remedy was offered by Jesuit priests who were regarded with great suspicion in Protestant Europe. In many countries with Protestant kings or queens, the entry of "Jesuit's bark," or "Jesuit powder," was forbidden. This led to an intolerable situation in which the Cinchona bark was banned in many countries in which malaria was causing great suffering among the population.

Jesuit bark soon became a much desired item among the smugglers of all nations. As it became obvious the curative power of the bark was real, necessity finally prevailed over prejudice. The medical community in all of Europe finally became alert to the miraculous power and curative compound found in the bark of a dozen Cinchona species, known then as "Fever Trees." The name was appropriate as the Cinchona bark in powdered form not only cured malaria but also reduced elevated temperatures during fevers of all kinds. The powdered bark of the Cinchona was soon being used as a primary treatment for malaria in other parts of the world as well.

The compound called "quinine," found in the bark of the Cinchona, is the active ingredient in the fight against malaria. This chemical compound's responsibility for the cure is something which was discovered in the year 1820 CE by two French scientists, Pierre

Joseph Pelletier and Jean Bienaime Caventou. These scientists identified the curative substance in the Cinchona bark as quinine. The compound called quinine is an alkaloid; it was named by these scientists after the indigenous word "quina," which is the native Indian word for bark.

Further chemical analysis of the Cinchona bark has revealed the presence of many other alkaloids. One of the compounds among these alkaloids is the substance called quinidine - mainly employed today in the treatment of abnormal rhythms of the heart, as well as in relieving muscular cramps, and as a remedy to treat severe headaches.

The demand for Cinchona during the 19th century in Europe and across the world increased to such an extent viable populations of the trees in South America were almost exhausted and the plant was threatened with extinction. European colonists made a series of attempts to cultivate different Cinchona species in their colonies in tropical regions of the world. Seeds and saplings of Cinchona trees were smuggled and even stolen in these efforts. In the end, the most successful Cinchona farmers were the Dutch colonists on the Indonesian island of Java, which turned out to have almost optimal conditions for growth of Cinchona trees. Currently the world's Cinchona supply mainly comes from central Africa and Indonesia, in addition to South America, its place of origin. Since the Cinchona is indigenous to the mountainous tropical regions of the South American continent, particularly the tropical area of Peru, it can also found in other countries in tropical South America.

Cultivation of the Cinchona tree is still big business, and Cinchona plantations can be found in Asian countries such as India as well as Indonesia. Cinchona is also cultivated in many parts of Africa. In these areas, the cultivation of the Cinchona is intensive and carried out on large commercial tree farms. Propagation of the Cinchona trees is done vegetatively from cuttings taken from the best trees late in the spring season to insure the quality of the bark. Harvesting the product requires the removal of trunk bark and bark from the branches and the root. The bark is only removed from six to eight year old trees. The collected bark is then dried in the sun. Annually, 10,000 tons of the bark is harvested from such farms.

Natural stands of Cinchona species have revived in many areas of the South American continent, a process which began when synthetic quinine was invented and the demand for bark decreased.

Quinine was synthesized because demand outstripped supply, especially when large numbers of non-native people were working or fighting in tropical areas, such as during the building of the Panama Canal and during the jungle fighting of WWII. Indonesia was in the hands of the Japanese and South America could not produce enough quinine for the Allied troops. It was simply necessary to get quinine from other sources and, as so often happens in wartime, resources were found to solve the problem. American chemists succeeded in synthesizing quinine in the year 1944.

Following the success of synthetic quinine, different quinine based medications like Chloroquine and Primaquine were subsequently used in treating malaria. These synthesized compounds were thought to be safer and more potent than the natural extracts of the bark. Eventually the malarial parasite began to develop resistance to the synthetic medication Chloroquine. This situation brought natural quinine out onto the shelves again for use in treating malaria. The effectiveness of Cinchona bark and quinine itself in the treatment of malaria has been questioned by recent evidence which shows the existence of certain sub-species of the malaria causing parasites are resistant to quinine. The discovery of these resistant variants of the parasite has sparked much debate about the real effectiveness of the whole plant over the synthetic varieties. This controversy did not lessen when the synthetic types began to lose potency against malaria, but became somewhat moot as new *Artemisia* based antimalarials became the treatment of choice.

Still, it is not time for quinine to retire. It has been used for fevers of all kinds and as a treatment for arthritis. It affects certain cancers. It is also a good influence on digestive problems and problems with heart arrhythmia.

Cinchona contains alkaloids (up to 15%), mainly quinoline alkaloids (quinine, quinidine), and also indole alkaloids (cinchonamine), tannins, bitter triterpenic glycosides (quinovin), quinic acid.

Of Oaks and Humans

An oak is a tree or shrub in the genus *Quercus* of which about 600 species exist on earth. The genus is native to the northern hemisphere, and includes deciduous and evergreen species. Oaks have great adaptability and versatility. Oaks are found in relatively warm and dry regions such as the coastal plans of California, in classic Mediterranean areas such as Greece, the Iberian Peninsula, and Italy, and in much of North American and Asia.

Oak trees have spirally arranged leaves, with a lobed margin in many species. Some have serrated leaves or whole leaves with a smooth margin. The flowers are catkins, produced by the tree in the spring. The fruit is a nut called an acorn, borne in a cup-like structure known as a cupule. Each acorn contains one seed and takes 6 to 18 months to mature, depending on species. The "live" oaks are distinguished for being evergreen, but are not actually a distinct group and instead are dispersed across the genus and the landscape.

Foliage and acorns in the autumn
©iStockphoto.com/Hans Laubel

Oaks were probably the first "bread" known to humanity, as areas where no grain was grown in ancient times are well stocked with ancient grinding stones, mortars, and pestles. Early human settlement often seemed to follow the oak forests as they ebbed and flowed across the planet during warm periods and icy ones.

The first digging sticks and plows were probably made from oak as well as many of the first handles for tools and hafts for weapons. Later in history the strength and elasticity of the Oak made it particularly valuable for house building and shipbuilding.

The phrase "Hearts of Oak," refers to the stately English manor houses, for the Englishman literally made his home from Oak. Many of the surviving old manor houses were constructed using huge oaken beams. Their walls were decorated with ornate, beautifully carved, oaken panels. Large solid Oak doors, sometimes bound with iron, secured the house from intruders and unwanted visitors.

The "Wooden Walls of England" is another old phrase associated with the Oak which refers to forts and castles constructed from Oak built around the coast to defend England from invasion, as well as the wooden sailing ships of England's Navy, the "ships made of oak," used in England's defense against the Spanish Armada. It is estimated it took a thousand mature oaks to build each ship. The Oaks of the Forest of Dean provided much of the material used for this, and Philip of Spain is said to have declared: "all the Oaks of the forest must be destroyed if victory is to be achieved." This he failed to do, but some two centuries later, so many of the Oaks had been felled and dispatched to naval dockyards for use in ships building, Lord Admiral Nelson drew up a special petition to the Crown advising the need to replant all the forests with Oaks.

Oak was also used in the construction of churches and cathedrals. The roof beams of Westminster Abbey are made from huge finely hewn trunks of Sessile Oak. In some areas, it was customary to plant a grove of Oak trees outside a new church or cathedral, so when a new roof was needed in a hundred years or so, the material would be well grown and close at hand.

In Sweden and Norway, Oak trees were not used by the common people. They were considered the property of the crown. Especially straight trees were marked and visited by royal foresters who would direct the royal woodcutters to trees marked for shipbuilding, especially tall, straight trees suitable for being made into masts.

As well as its strength for building purposes, the Oak is much prized for the beauty of its grain and texture, and the richness of its coloring after polishing. As such it has always been a favorite wood of carpenters and cabinetmakers for use in paneling, doors, and furniture. Beautiful cupboards, chests, tables, and chairs were made of Oak. Due to the wood's durability many of these have survived down through the centuries. Initially pale brown or golden in color, Oak wood darkens with age and its grain becomes more distinct.

Other uses of Oak were the fighting clubs of ancient man, the hammers, spears, and long boats of the Vikings, and the hafts of daggers and knives made from its roots, a practice which was supposed to give a warrior great strength in battle. Barrels and casks were also made from Oak and used to store liqueur, wines, and spirits, as oak wood is impervious to the effects of alcohol and will not split or become soft. Coffins were made of Oak by using large sections of the trunk of the trees. These sections were split lengthwise and hollowed out to contain the body, but this was only done for state funerals or people of great stature and importance. The shrine of Edward the Confessor in Westminster Abbey is of marble, but his coffin, made in the year 1510 CE, is of Oak, which has outlasted the changes of some 700 years.

It is for reasons such as these the Oak is called the "frame of civilization."

Frankincense, Myrrh, and Balm of Gilead—the incense trees

Frankincense, also called olibanum (Arabic: lubban, Hebrew: levonah), is an aromatic resin obtained from trees of the genus Boswellia. Frankincense is tapped from the desert-adapted, hardy Boswellia tree which grows in rocky and arid lands around the Red Sea and the Arabia Sea. There are numerous species and varieties of frankincense trees, each producing a slightly different type and color of resin. Differences in soil and climate and available water create even more diversity of the resin, even within the same species. These trees are found on the Horn of Africa, on the coasts of Yemen, and the Arab Emirates. Some survive not by the sparse rainfall of these desolate regions, but by the fog which condenses on the complex and crinkly leaves and runs down their trunks to water the earth about them.

There are four main species of Boswellia which produce true frankincense, particularly *Boswellia sacra* (syn. *B. carteri*, *B. thurifera*), *Boswellia frereana*, *Boswellia serrata* and *Boswellia papyrifera*.

The Frankincense resin is used in incense, as well as in perfumes. Each type of Frankincense resin is available in various grades. The grades depend on the time of harvesting. The resin is hand-sorted for quality. The sap is collected by incising the bark and allowing the exuded resins to ooze out and harden on the trunk of the tree. These hardened resins are called tears. Frankincense trees grow in environments so unforgiving they sometimes grow directly out of cliffs and rock faces. The means of initial attachment to the stone is thought to be the stickiness of seeds of the Buseraceae family but this is accompanied by a burl-like twisting and bulbous disk-like swelling of the trunk which clings suction cup-like to the face of the rocks. This odd growth at the base of the tree prevents it from being torn away from the rock during the violent sandstorms which frequent the regions where the trees grow. This feature is slight, but present, in trees grown in rocky soil or gravel and can be seen by examining the trunk just above ground level. The tears from the hardy rock-grown survivors are considered superior due to their stronger and sweeter fragrance.

Frankincense tree in the mountains of Oman
©iStockphoto.com/Maros Markovic

The trees start producing harvestable quantities of resin when they are about 10 years old but even a leaf removed from a young tree will yield a drop of milky sap at the leaf base. The raw sap smells pleasantly of flowers, but its true fragrance is not revealed until the dried sap is burned. Tapping is done 2 to 3 times a year with the final taps producing the best tears due to their higher aromatic terpene, sesquiterpene and diterpene content. The thicker cloudier resins are considered the best in quality. Dhofari frankincense (from *Boswellia sacra*) is said to be the best in the world from a medicinal point of view (frankincense is a powerful anti-inflammatory), although fine resin for incense is produced more extensively in Yemen and along the northern coast of Somalia.

Recent studies have indicated frankincense tree populations are declining due to unwise and exploitative tapping rates. Trees weakened by tapping are more likely to be damaged by storms and less resistant to drought. Heavily tapped trees have been found to produce seeds with very low germination rates while seeds of trees which have not been tapped germinate at much higher rates.

Frankincense has been traded on the Arabian Peninsula, in the Middle East, and in North Africa for more than 5,000 years. A mural depicting sacks of frankincense traded from the Land of Punt and workers gathering sap from frankincense trees adorns the walls of the temple of ancient Egyptian Queen Hatshepsut, who died in 1458 BCE. The mural is clear enough to identify the species of frankincense trees depicted, a plant which is now called *Boswellia papyrifera*.

Frankincense was reintroduced to Europe by the French Crusaders. Although it is known as “frankincense” in most of Europe, the resin is also known as olibanum, which is derived from the Arabic al-luban (roughly translated: “that which results from milking”), a reference to the milky sap tapped from the Boswellia tree. Some have also postulated the name comes from the Arabic term for "Oil of Lebanon" since Lebanon was the place where the resin was sold and traded with Europeans.

Much earlier this resin was mentioned in the Bible in Exodus 30:34, as levonah, meaning "white."

The Greek historian Herodotus was familiar with Frankincense and knew it was harvested from trees in southern Arabia. The resin is also mentioned by Theophrastus and by Pliny the Elder in his *Naturalis Historia* and by several of the ancient Greek physicians, who considered it something of a cure-all and included it in many medicines.

Currently the properties of Frankincense are being tested against inflammatory disease of all kinds, sports injuries, and inflammation of the skin and cancer.

Myrrh (from Morr or "bitter" in Hebrew and Murr in Arabic) is the modern name for a fragrant sap, a reddish-brown resinous material, which is the dried sap of a number of different trees, but primarily from *Commiphora myrrha*, a tree which is native to Yemen, Somalia, and the eastern parts of Ethiopia. The sap of a number of other *Commiphora* species is also known as "myrrh," adding to the confusion and mystery which surrounds this substance.

Good quality myrrh can be identified through the color and clarity of the resin. The best method of judging the resin's quality is by judging the stickiness of freshly broken fragments and the scent to determine the fragrant-oil content of the myrrh resin. The scent of raw myrrh resin and its essential oil is sharp, pleasant, somewhat bitter, and can be roughly described as being resinous. When burned, it produces a smoke which is heavy, bitter, and many with a layered scent, which may be accented with a slight floral sweetness. Unlike most other resins, myrrh expands and "blooms" when burned instead of melting or liquefying into oil.

The powerful smell of myrrh can also be used in mixtures of incense, to provide an anchoring element to the overall scent. It has been used as an additive to wine, both for taste and for medicinal purposes. It is considered an effective topical ointment for skin eruptions and diseases. It is also used in various perfumes, toothpastes, lotions, and other modern toiletries.

Myrrh was used for embalming in ancient times, as a penitential incense at funerals and cremations, and as a sacrificial material. The holy oil traditionally used by the Orthodox Church in the Middle East for performing sacraments of chrismation and unction is traditionally scented with myrrh, and receiving either of these sacraments is commonly referred to as "receiving the Myrrh."

The Ancient Egyptians imported large amounts of myrrh as far back as 3000 BCE. They used it to embalm the dead, as an antiseptic, and burned it for religious sacrifice. Myrrh has been traded throughout the Middle East at least since 1500 BCE. The origins of myrrh are traced to the Arabian Peninsula and the legendary kingdom of Sheba. The collection of the gum resins for export was initiated in Sheba. According to Herodotus (5th century BCE), "Arabia is the only country which produces myrrh, frankincense, cassia, and cinnamon." Diodorus Siculus writes, in the second half of the first century BC, "all of Arabia exudes a most delicate fragrance; even the seamen passing by Arabia can smell the strong fragrance which gives health and vigor." Other comments include those about the fragrance the "dhows" loaded with frankincense and myrrh left behind them as they sailed. These comments were exaggerations, of course, as the resin being shipped was well packed to save it from sea spray and therefore scentless. Cinnamon and cassia probably came from India at that time and were only exported through the legendary kingdom, which was too arid for either tree to thrive. The myrrh and frankincense trade route reached Jerusalem and Egypt from what is today Oman (known in the past as the Dhofar region), and from Yemen, following the Red Sea coast of Arabia and may have gone as far as Timbuktu to the west and China to the east. The three Wise Men (Magi) carrying and delivering the myrrh, gold, and frankincense, for the baby Jesus came from the Arabian peninsula, perhaps from the legendary kingdom itself or one of its lost colonies on the Horn of Africa.

Myrrh Trees
©iStockphoto.com/Vladimir Melnik

In ancient history myrrh was used as a constituent of perfumes and incense, as medicine, and was worth more than its weight in gold. The Greek word for myrrh, *μύρον*, came to be synonymous with the word for "perfume." Today myrrh is also valued for its antimicrobial properties especially in the context of skin diseases.

In Ancient Rome myrrh was priced at five times as much as frankincense, though the latter was more in use. Myrrh was burned in ancient Roman funerals to mask the smell emanating from burning bodies. It was said the Roman Emperor Nero burned a whole year's harvest of myrrh at the funeral of his wife Poppaea. Pliny the Elder refers to myrrh as being one of the ingredients of perfumes, and specifically the "Royal Perfume" of the Parthians. He also says myrrh was used to fumigate wine jars before bottling and sometimes added to wine. Archeologists have found at least two ostraca from Malkata (from New Kingdom Egypt, ca. 1390 to 1350 BCE) which were lined with a shiny black or dark brown deposit analysis showed to be chemically closest to myrrh. The Romans were known to use myrrh as a premier additive to wine which promoted good digestion. They also used myrrh to cure mouth ulcers and pack bad teeth.

Balm of Gilead seems to have been a special species of myrrh. It may have been called "afarsemon" by the Hebrews or by the more common biblical name "tzori Giladi." The Gilead region is an area currently shared between Israel and Jordan. The tree no longer survives in this area, which is much drier now than it was in Biblical times. Considered a medicine of last resort and given in a cup of wine to the very ill, this type of myrrh was worth one hundred times as much as frankincense. The tree was cultivated in only three places in the ancient world, one of them at Ein Gedi where the floor of an ancient synagogue warns, "Cursed be he who reveals the secret of our community." The floor of the Ein Gedi synagogue is on display at the Rockefeller museum in Jerusalem where the inscription in Aramaic is clearly visible.

The production of resin, the distilling of perfumes and medicines, and the propagation and cultivation of the trees was a set of important trade secrets which protected the Jewish inhabitants of Ein Gedi for centuries from successive waves of conquerors and invaders after failure of the revolt against Rome in 70 CE. The fortunate inhabitants of the three centers of Balm of Gilead cultivation were neither killed nor taken into slavery as they were needed to tend the trees and keep up the production of the valuable products.

The Romans and the Sabeans fought a war over who would control the trade in the resin from the above mentioned trees. The Sabeans lost and the Kingdom of Sheba with its many African colonies and great networks of dams and canals was deprived of vital revenue and went into decline.

Frankincense, Myrrh, and Balm of Gilead are all being tested for their medicinal properties in modern times as they have proved to be anti-inflammatories and germicides and to have anti-cancer properties in addition to their other qualities. They may become very important to modern medicine in the near future. However, in the ancient world these trees were important enough to set off trade wars, initiate conquests and save defeated and despised people from the worst intentions of their conquerors.

Coffee, Tea and Cocoa: the Engines of Trade

These trees produce products so sought after they sparked exploration, caused plantations to be developed in the far corners of the earth, and lead to populations being uprooted and moved to tend to them.

Coffee

Several species of shrub of the genus Coffea produce the berries from which coffee is extracted. The two main species commercially cultivated are *Coffea canephora* (predominantly a form known as 'robusta') and *Coffea arabica. C. arabica*, the most highly regarded species, is native to the southwestern highlands of Ethiopia and plateaus in southeastern Sudan and possibly one mountainous area in northern Kenya. *C. canephora* is native to western and central sub-Saharan Africa. Less common species are *Coffea liberica, C. excelsa, C. stenophylla, C. mauritiana*, and *C. racemosa*. They are used locally and sometimes in the breeding process for new coffee varieties.

Coffee tree with ripe and green beans
©iStockphoto.com/Ericka Norman

All coffee plants are classified in the large family Rubiaceae. They are evergreen shrubs or small trees which may grow 5 m (15 ft) tall when left to obtain their natural height and shape. The leaves are dark green and glossy, usually 10 to 15 cm (4 to 6 in) long and 6 cm (2.4 in) wide. The flowers are axillary, and clusters of fragrant white flowers bloom simultaneously and are followed by oval berries of about 1.5 cm (0.6 in). The berries are green when immature, but ripen to yellow, then crimson, before turning black on drying. Each berry usually contains two seeds, but 5 to 10% of the berries have only one. These are called peaberries. Berries ripen in seven to nine months.

Coffea arabica is predominantly self-pollinating, and as a result the seedlings are generally uniform and vary little from their parents. In contrast, *Coffea canephora*, *C. excelsa* and *C. liberica* are self-sterile and require outcrossing. This means useful forms and hybrids must be propagated vegetatively to insure quality and uniformity of harvest. The rooting of cuttings into cloned trees, grafting valuable scions on less valuable rootstocks, and budding on strong seedling trees are the usual methods of vegetative propagation. These methods can be problematic and increase the cost of coffee cultivation. On the other hand, in the need for in-species diversity there is great scope for experimentation in search of potential new varieties. This genetic richness underpins the ability to breed coffee trees to meet the challenges of the day.

Some types of coffee are propagated by seeds. The traditional method of planting coffee is to put many seeds in each hole at the beginning of the rainy season. Half are eliminated naturally, then to select the strongest from those which sprout. With coffee types which require vegetative propagation the trees are propagated in nurseries where the scions of the desired coffee types are carefully grafted or budded on strong, well adapted rootstocks. A more effective method of growing coffee, used in Brazil, is to raise seedlings in nurseries for up to two years then plant out trees which are large enough and strong enough to survive. Coffee is often intercropped with food crops, such as corn, beans, or rice during the first few years of cultivation or planted with arboreal companions called "nurse trees" which shade and protect the young plants.

Coffee is a brewed drink prepared from roasted seeds of the coffee plant, commonly called coffee beans or coffee berries. Once ripe, coffee berries are picked, sorted, and dried. The seeds are then roasted to varying degrees, depending on the desired flavor. They are then ground and brewed to create coffee. Coffee can be prepared and presented in a variety of ways with various types of perking and filtering, and various additions of everything from sugar, to spices, to chicory, and milk.

The energizing effect of the coffee bean plant is thought to have been discovered in Yemen in Arabia and in the northeast region of Ethiopia. The cultivation of coffee first expanded in the Arab world, starting apparently in the more temperate areas of Yemen. Early coffee cultivation was also practiced in the hills of Ethiopia. By the late fifteenth century, coffee drinking was widespread in the Muslim world. From the Muslim world, coffee spread to Italy, Spain, then to the rest of Europe, to Indonesia, Asia, and to the Americas. Coffee has played an important role in the development of trade worldwide. In Africa and Yemen, it is used in religious ceremonies. It was banned in Ottoman Turkey during the 17th century for political reasons, and was associated with rebellious political and intellectual activities in Europe.

An important export commodity, coffee was the top agricultural export for twelve countries in 2004 CE and it was the world's seventh-largest legal agricultural export by value in 2005 CE. Some controversy is associated with coffee cultivation and its impact on the environment as it requires very specific conditions and is usually grown in orchards formats which are not always ecologically friendly. Coffee plants tend to be vulnerable to a wide variety of insect pests and crop protective sprays have limited power to protect them. Coffee leaf rust is the main coffee disease, sparking the development of *Coffea robusta* from *C. canephora* and the other coffee types simply because it was less vulnerable to the rust, the many other coffee pests, and grows at a lower altitude, therefore extending the range in which coffee can be profitably grown. Currently *Coffea robusta* accounts for about twenty five percent of coffee cultivation but *Coffea arabica* remains the favorite.

There is also discussion as to whether the general influence on people's health is negative or positive with a good deal of evidence to the yea and the nay. Due to its caffeine content, coffee can have a stimulating effect. It also seems to be good for cardiac health and can be an effective laxative.

What is a fact is coffee is one of the most valuable agricultural products in the world. Coffee beans are grown on trees in over seventy countries. Green unroasted coffee is one of the most traded agricultural commodities in the world. Today, coffee is one of the most popular beverages worldwide

Tea

Tea is the second engine of commerce with whole fleets from Europe dispatched to the East to acquire tea. Fortunes were made and lost depending on the quality and quantity of the tea they brought back.

Camellia sinensis is the species of plant whose leaves and leaf buds are used to produce tea. It is of the genus Camellia, a genus of flowering plants in the family Theaceae. White tea, green tea, oolong, smoked tea, and black tea are all harvested from this species, but are picked, processed and cured differently to attain different levels of oxidation. There are two major varieties which characterize this species, the small-leaved Chinese variety plant (*C. sinensis sinensis*) and the large-leaved Assamese plant (*C. sinensis assamica*).

Still picked by hand, fresh leaves contain 4% caffeine. The young, light green leaves are the ones harvested for tea production. They are shiny and slick on the top side of the leaf with short white fuzz on the underside. Older leaves are deeper green and have a much higher tannin content. Different ages of leaves produce different qualities of tea, since their chemical compositions change as the leaves mature. The tip and the first two to three leaves are harvested and then dried and processed. Tea can be picked every two weeks in tropical areas and is a labor intensive crop with very little successful mechanization.

Camellia sinensis is mostly cultivated in tropical and sub-tropical climates, in areas with at approx. 130 cm of rainfall a year as tea plants prefer moist, rich soil. Many high quality teas are grown at high elevations, up to 1500 meters as the plants grow more slowly and acquire a better, deeper, but less bitter flavor. Tea plants will grow into a medium sized tree if left undisturbed, but plantation plants are pruned to waist height for easy harvesting.

Tea drinking originated in China and the word tea is derived from "t'e" of a Chinese dialect. Legend has it, tea was discovered by the second emperor of China when leaves blew into his morning cup of hot water. This supposedly occurred in 2727 BCE. He recommended it to his court and military saying the drink was invigorating and cleared the head. The first authenticated reference to tea was made in an ancient Chinese dictionary revised by Kuo P'o, a celebrated Chinese scholar in 350 CE. This work refers to tea as "Erh Ya." At that time, a medicinal decoction was made by boiling tea leaves. The use of tea as a beverage became more common towards the close of the sixth century. During the next centuries, tea gained enormous popularity. The first exclusive book on tea, *Ch'a Ching* meaning "Tea classic" by the Chinese tea expert, Lu Yu, was published in 780 CE. In the book Lu Yu described various kinds of tea, the cultivation of tea trees, and curing and storing of tea leaves in China.

Tea drinking next spread to Japan with Buddhist monks who came to China to study, became acquainted with the tea plants, and took seeds and saplings back to Japan with them in 593 CE. In 648 CE, a Japanese monk named Gyoki, planted the first tea bushes in 49 Buddhist temple gardens. Tea in Japan was rare and expensive, enjoyed mostly by high priests and the aristocracy.

Apart from Japan, tea drinking did not spread to other parts of the world until about the middle of the seventeenth century. The Dutch introduced it to Europe. In Cantonese, tea is known as "Ch'a" and this is the name by which tea came to be known in India, Russia, Iran, and the Middle East. The opening of a sea route to India and the East by the Portuguese in 1497 CE allowed for large-scale trading between Europe and the Oriental countries. Other European nations soon followed the Portuguese in establishing trade centers in different countries of the East. The Dutch in Java established one such trade depot. They bought tea from Japan, packed it in Java in waterproof barrels, and the first consignment of tea was transshipped from Java to Europe in 1610 CE. This marked the beginning of the extremely lucrative tea trade between Europe and the East. The Dutch dominated the tea trade for more than a century before finally yielding to the British. China was the sole supplier of commercial quantities tea to Europe till the middle of the nineteenth century. This was accomplished almost exclusively by ship, though some supplies still trickled westward over the land routes of trade through Mongolia and Turkey.

Tea-drinking gained popularity among the affluent sections in Europe within fifty years of its first introduction into the continent. It was, however, extremely expensive and the tea preparation equipment of the day reflect its dearness in the form of locked caddies, tea chests which look like safe deposit boxes, and strainers designed to squeeze the last drops of liquid out of the precious leaves. In about another 100 years it became a beverage of daily enjoyments in a large part of Europe and Britain. Tea also became popular in America, which was then a British colony, and was also introduced to Canada. By 1670 CE the Massachusetts colony was drinking black tea. So popular did the drink become, a tax on tea in the colonies was the spark which set off the American Revolution in 1776 CE.

The discovery of the Assam tea plant broke the Chinese monopoly on cultivated forms of tea. It is attributed to Robert Bruce who is supposed to have seen the plant growing wild in some hills near Rangpur (near present Sibsagar), then the capital of Assam, during his visit in 1823 CE on a trading mission. He died before he could follow up on his plans to investigate and develop the plant. Instead, his brother met and came to an agreement with the local chief who supplied him some tea plants and seeds.

In 1834 CE, the then Governor General of India, Lord William Bentinck, appointed a Tea Committee to advise on feasibility of commercial tea cultivation in India. The committee issued a circular asking for information on areas suitable for tea cultivation and sent its secretary to procure tea seeds, plants, and workers from China. The Commissioner of Assam however made a strong case in favor of tea cultivation in Assam where tea plants were growing wild in the forest. He also collected complete specimens of the local plants and forwarded them to the Government Botanical Gardens in Calcutta. When the specimens were positively identified as tea plants, the Tea Committee recommended the indigenous plant should be cultivated for commercial purposes, if possible, instead of the Chinese plants.

Tea plantation in Sri Lanka
©iStockphoto.com/Erkki Tamsalu

This was a very important development because not only did it establish the worth of the Assam tea plant but determined the future course of tea cultivation throughout the world. Soon Assam type plants were being grown in many places in India, Thailand and Sri Lanka. Today, more tea is made from the Assam type of plants than from the China type and tea in its many forms has become the most widely quaffed beverage on the planet.

Cocoa, the Drink of the Gods

The source of chocolate in all its forms, the Cocoa tree, has delighted millions around the world by producing the basic material of many confections, cakes, and beverages. It was also a sought after trade item like coffee and tea which initially cost so much only royalty could partake of it.

The Cocoa tree (*Theobroma Cacao*) grows in the warm and humid equatorial belt very close to the equator. Although the origins of the tree are disputed, it can be traced to the tropical regions of Venezuela, Honduras, and Mexico. There is a good deal of proof which points to the ultimate origin of cocoa and chocolate in the Ulúa valley in Honduras. Today cocoa is cultivated globally, albeit in a narrow belt around the equator, in carefully grown plantations in the tropical rainforests of Africa, Asia, and Latin America.

The perfect environment for the cocoa tree is the moist tropical heat of the equatorial forests and their shaded, well-watered soils. Young cocoa trees only thrive in tropical temperatures within the protective shadow of tall plants like bananas or palm trees which shelter them from the sun and wind. Direct sunlight and dryness are enemies of the delicate young trees.

The trees begin to bear in the fifth year of life and will continue for another twenty years to bloom and produce the pods of cocoa. The cocoa tree flowers in two cycles of 6 months, in other words twice a year. Thousands of white (female) and pink (male) tiny flowers with five petals adorn the stem and branches of the cocoa tree. Only a few hundred will be fertilized and no more than forty per tree will develop into cocoa pods. Cocoa pods resemble long green melons.

After 6 months the cocoa pods are full-grown and have changed color from green to yellow-orange. The pods are carefully harvested by the plantation workers. The cocoa pods ripen for a few days after the harvest. The outer peel is opened using long knives with care taken not to damage the beans. There are two harvests per year. The beans are then dried, aged, sorted, graded, and shipped to where they will be processed.

Cocoa tree with fruits
©iStockphoto.com/Kseniya Ragozina

Cocoa Tree Varieties

There are three different varieties of cocoa tree. The descendants in the plantations today are usually of these varieties or cultivated hybrids of these varieties, each with their own particular characteristics:

Criollo is known as the best of the cocoa trees. The tree produces pods with a very thin peel. The cocoa itself has a very pale color and a strong but refined aroma. This variety produces small harvests and is also sensitive to the weather.

Forastero is a stronger type of tree which is easier to cultivate and produces larger yields. The cocoa pods have a thicker peel and a coarser, stronger aroma as well as a rich dark color. Cocoa from the Forastero beans is often called bulk cocoa because it gives chocolate a typical recognizable basic aroma. This cocoa forms the basic ingredient in most chocolates and can often account for 80% of the cocoa mixture.

Trinitario is a cross of both types of trees and has characteristics of both of the parent plants. It has a strong but relatively refined aroma, and moreover, is very easy to cultivate.

A Very Condensed Commercial History of Chocolate

Cocoa began as the base of a bitter drink, the tree first cultivated by Olmec Native Americans in 1500 BCE. Around 400 BCE the Mayan Native Americans begin to cultivate the tree. From 250 to 900 CE the consumption of cocoa beans was restricted to the Mayan society's elite, in the form of an unsweetened cocoa drink made from the ground beans. Around approximately 600 CE, the Mayans migrated into northern regions of South America establishing the earliest known cocoa plantations in the Yucatan, as well as introducing many of their other domesticated crops. In the 14th Century the drink became popular among the Aztec upper classes who adopted the cocoa beverage from the Mayans and took cocoa beans as tribute. The Aztecs called it "xocalatl" meaning warm or bitter liquid.

In 1502 CE, Columbus encountered a great Mayan trading canoe carrying cocoa beans as cargo. He was puzzled as to the reason for the beans value. In 1519 CE, Spanish explorer Hernando Cortez recorded the cocoa usage in the court of Emperor Montezuma and declared the drink was wholesome and pleasing unlike many other Aztec foods which he said were devilish. In 1544 CE, Dominican friars took a delegation of Mayans to visit Prince Philip of Spain. The Mayans brought gift jars of beaten cocoa, mixed and ready to drink. Spain and Portugal established a monopoly on cocoa beans and did not export the beloved drink to the rest of Europe for a hundred years. It was during the 16th Century the Spanish began to add cane sugar and flavorings such as vanilla to sweeten cocoa beverages. In 1585 CE, the first official shipments of cocoa beans began arriving in Seville from Vera Cruz, Mexico.

In 1657 CE, the first chocolate house was opened in London by a Frenchman. The shop was called "The Coffee Mill and Tobacco Roll." Expensive at 15 shillings per pound, chocolate was considered a beverage for the moneyed classes. In 1674 CE, eating solid chocolate was introduced in the form of chocolate rolls and cakes, served in special shops and chocolate emporiums.

In 1730 CE, cocoa beans had dropped in price from $3 per pound and became more generally available. In 1732 CE, French inventor, Monsieur Dubuisson, invented a table mill for grinding cocoa beans which made a very fine powder suitable for baking and brewing. In 1753 CE, Swedish naturalist, Carolus Linnaeus, a great fan of hot chocolate, was dissatisfied with the word "cocoa," so he renamed it "Theobroma," which is from the Greek for "food of the gods." In 1765 CE, chocolate was introduced to the United States when Irish chocolate-maker John Hanan imported cocoa beans from the West Indies into Dorchester, Massachusetts, to refine them with the help of American Dr. James Baker. The pair soon after built America's first chocolate mill and by 1780 CE, the mill was making the famous BAKER'S® chocolate sold in large slabs for baking, and making candy, and beverages. In 1795 CE, Dr. Joseph Fry of Bristol, England, employed a steam engine for grinding cocoa beans, an invention which led to the manufacture of chocolate on a large scale.

In 1828 CE, the invention of the cocoa press, by Conrad Van Houten, helped reduce the price and improve the quality of chocolate by squeezing out some of the cocoa butter and giving the beverage a smoother consistency. Conrad Van Houten patented his invention in Amsterdam and his process became known as "Dutching." The process allowed the cocoa to mix better with water. In 1830 CE, a form of solid eating chocolate was developed by Joseph Fry & Sons, a British chocolate maker. In 1847 CE, Joseph Fry & Son discovered a way to mix some of the cocoa butter back into the "Dutched" chocolate, and added sugar, creating a paste which could be molded and would set when cooled. This was the first modern chocolate bar. In 1868 CE, John Cadbury marketed the first boxes of chocolate candies. In 1876 CE, Daniel Peter of Vevey, Switzerland, experimented for years before finally inventing a way of making the milder and sweeter "milk chocolate." Daniel Peter and Henri Nestlé soon combined forces to form the Nestlé Company. In 1879 CE, Rodolphe Lindt of Berne, Switzerland, produced a smooth and creamy chocolate which melted on the tongue. He invented the "conching" machine. This machine heated and rolled chocolate in order to refine it. After chocolate had been conched for seventy-two hours and had cocoa butter added to it, it was then possible to create chocolate "fondant" and other creamy forms of chocolate.

The first known published recipe for chocolate brownies appeared in the Sears and Roebuck Catalogue in 1910. Soon after, a Canadian by the name of Arthur Ganong marketed the first nickel chocolate bar. William Cadbury urged several English and American companies to join him in refusing to buy cocoa beans from plantations with poor labor conditions, marketing an early form of "fair trade" chocolate. In 1913 CE Swiss confectionist Jules Sechaud of Montreux introduced a machine process for manufacturing filled chocolates. In 1926 CE, Belgian chocolatier, Joseph Draps, starts the Godiva Company to compete with Hershey's and Nestle's American market.

And then: chocolate becomes the most sought after and important confectionary in the world, one of the most prized products of the20th century. Not bad for a tree which originated in one valley, high in the mountains of Central America.

And Finally, Kola, the Bitter Stimulating Nut of Africa

Kola nut trees are most common in Western Africa and the Atlantic coast area of Central Africa. Kola nuts (or cola nuts) are the seed pods of these various large evergreen trees which grow in rich, well-watered soil. Sterculiaceae *Cola vera* is the scientific name of the most common species. It is related to the South American genus *Therobroma* or cocoa. They are evergreen trees, growing to 20 m tall, with glossy ovoid leaves up to 30 cm long. The tree can obtain a height of up to 60 feet tall (about 18 meters). The fruit is a heavy pod with a rough outer layer and with many sections within.

The use of the kola nut, as with the use of coffee, cocoa, and tea has ancient origins. It was chewed in antiquity to ease hunger pains, give energy on the hunt, and to help people on long journeys. It was not forbidden to African Muslims, as was beer and wine, and so became popular with them. In many West African cultures it is used in a religious context or chewed to restore vitality. Kola nuts are an important part of the traditional spiritual practice of the Yoruba culture and religion in West Africa. Kola nuts are used as religious object, gifts, good luck charms, and as a sacred offering during prayers, ancestor veneration, and significant life events, such as births, naming ceremonies, weddings, and funerals.

Brightly colored kola nuts are a common sight in African street markets and shops in cities and villages. They are often sold by street vendors at bus and train depots and in village markets as well. On a train or bus, a traveler with a kola nut will often offer a piece to the other passengers much like a stick of gum or a piece of candy might be offered in the US.

Kola nuts are consumed by breaking the pod open and dividing the sections which fit closely inside the pod into pieces. Then the pieces are chewed one by one. Most people find the taste very bitter. The stimulative effect is similar to a very strong cup of coffee. Kola nuts are produced commercially in the African and American tropics. In their raw form they are rather hard to find outside these areas, although cola extracts are currently exported all over the world.

Extracts of these bitter nuts are often added to energy drinks. Kola nuts are best known outside of Africa as an ingredient in cola beverages. There is some evidence the first kola beverage was made by Western Africans who mixed water with dried or fermented kola nuts, strained the liquid, and then added cane juice. Today, homemade cola drinks are very rare in Africa, though store-bought cola drinks and homemade beers of all kinds including cola beer are very popular.

Cola acuminate nuts and glass of cola drink
©iStockphoto.com/Jamie Watson

Commercially produced cola drinks were developed in the late 1800s, when chemists and inventors and manufacturers began to use kola nuts in carbonated drinks and tonics. The most famous of these is Coca-Cola®, which has become a truly global beverage and is only lightly less popular than tea, coffee, wine, and beer.

Recently, kola nuts and kola nut extract have become popular in Europe and North America as natural medicines. Kola nuts are often used to treat whooping cough and asthma. The caffeine present acts to expand the bronchial air passages and suppress the cough.

Kola nuts are valued mainly for their stimulant qualities. They were once commonly used in soft drinks such as Coca-Cola and Pepsi, but have now been replaced with artificial ingredients. The kola extracts have effects similar to other xanthine-containing plant products such as chocolate, tea, coffee, guarana berries, and yerba maté. They have stimulant effects on the central nervous system, the circulatory system and the heart. They increase body temperature, raise blood pressure, and elevate respiratory rate. Effects may last for hours. In medicine the refined extract is used as a cardiac and central nervous system stimulant.

Imagine the world without these trees and products.

Would the ships of the explorers even have been built without the sturdy oak?

Would people have been able to live in South America, in southern USA and many other places without a reliable cure for malaria?

The incense trees, precious balms and perfumes from the past may have an even more interesting future......

What would the world be like without tea, coffee, chocolate, or cola?

Surely it would be a different world than the world which exists today.

Chapter 15
Where Are the Trees?

Trees in the developed world are often conspicuous by their very absence.

City Trees

Trees which once lined avenues and made urban centers bearable have mostly disappeared from large cities. Their shade, their ability to absorb noise, and the rest they gave the urban eye is very much missed because cities have cut down existing trees and do not plant new ones. The reasons given for denuding the cities of trees are unconvincing:

Trees drop leaves and dirty streets.

People run into them with automobiles and sue the municipalities.

Tree branches get in the way of electric lines.

Traditional tree-lined street, USA
©iStockphoto.com/DNY59

It is for reasons such as these "modern" cities are for the most part "treeless" cities.

Trees are the first line of protection for the pedestrian, shielding the pedestrian from weather, providing shade and putting a barrier between the person on foot and the rushing traffic. However, the pedestrian's comfort is no longer considered important because the pedestrian is no longer important. What is important is the flow of vehicle traffic. Therefore, trees have vanished and square corners have been shaved round to allow vehicles to swoop around them without slowing down. It has become steadily more dangerous and more unpleasant to walk in many urban areas, even in places designed for tourists.

The result of this eerie urban "treelessness" is a sort of artificial desert where one sees endless naked concrete streets and buildings, outsized commercial signs, and light glittering mercilessly on the glass and chrome of thousands of automobiles. The involuntary signs of pain, registering on the faces of volunteers during stress testing, when shown a typical urban area, must surely be an unvoiced plea for change.

As their eyes dart from point to point of the bleak vista, it can be assumed they are looking for something pleasant to rest their gaze on. They do not find it and look at the scenes of parking lots, franchises, and billboards for the minimum amount of time allowed by the study. The same group of volunteers shown shady meadows, forests, orchards, and even natural deserts full of Joshua trees and cacti, tend to look at these photos for the maximum amount of time—obviously finding something in the photos which is both pleasant and interesting. Their faces relax and their gaze lingers.

If this is a normal reaction and we can make people feel better just by lining the city streets with trees, there is so much benefit for so little investment. Where are the trees? Why aren't we planting them?

Suburban Trees

In the suburbs one rarely sees a well grown tree, at least one large enough and strong enough to give good shade or provide protection from the wind. Vigorous trees are pruned into wisps so as not to interfere with infrastructure or the neighbor's view. Backyard fruit trees are a thing of the past in most locales for sanitary reasons. Suburban lawns are usually "landscaped by shrubs," useless ones at that.

A road lined with trees
©iStockphoto.com/Giorgio Fochesato

Why should lawns and unproductive ornamental plants be the standard plantings of suburban areas? There should be room for productive fruit trees, herbs, berry bushes, copses, windbreaks, picnic groves, woodlots, and maybe a large spreading tree over the post office or the local library, replacing the lost chestnuts of North American villages.

Rural Trees

A hundred years ago when there were still family farms in the US and Canada, almost every farm had a woodlot and an orchard. The orchard produced the family's fruit and the woodlot was a source of logs and kindling, wood chips, and forest products which contributed directly to the household economy. A four acre woodlot could fill the energy needs of a 40 to 60 acre farm. These wooded areas were also places where moisture was stored naturally, refuges for wild creatures, islands of shade in the landscape.

One of the best examples of ecological farming described in Michael Pollen's book *The Omnivore's Dilemma*, Penguin; Reprint edition (August 28, 2007), is the Polyface farm owned by Joe Salatin, which is discussed in chapter 8 of Pollen's book. Here, in the chapter entitled "All Flesh Is Grass," Author Pollen waxes lyrical over descriptions of the farm, the succession of crops and animals, the farm's great diversity of products; over how much is utilized and how little is wasted.

Joe Salatin's farming methods are explained by him in his own book, *Pastured Poultry Profits*, Polyface; Reissue edition (July 1996). However, the beginning of this sustainable, productive, and ecologically sound enterprise was a vigorous tree planting program initiated by Joe Salatin's father. Salatin Senior planted the acres of trees which transformed worked-out, eroded, and desolate land into forest, cooling the area and shielding it from more erosion while creating a microclimate his son turned into a food production system based on grass.

Woodlot
©iStockphoto.com/Tomasz Budnik

Anti-Tree Developments in the World

The practice of replacing the trees unfortunately is the exception rather than the rule when rural lands are developed. In the "development" of rural lands, the trees are usually the first thing to go, logged off or simply pushed up by a bulldozer and burned. Later, spindly, non-native plants will be placed at useless intervals along the gently curving characterless streets which have replaced productive farmlands all over North America. This deliberate destruction is tragically the official policy in many parts of the world.

In Australia land was claimed by "clearing the bush"—this means removing all vestiges of natural vegetation. The result was not a healthy replacement set of transplanted European cultivars but a degraded landscape, in many areas increasingly parched and saline. Australia's flora is adapted to the continent's ancient and marginally fertile soils, its saline water tables (with precious fresh rainwater floating precariously atop the salty water) and its extreme winds and temperatures. Australia's flora protected the fragile soil in a way which no imported plant could ever do. Many of the island continent's serious ecological woes stem from the bush clearing policy and now some organizations are working to put the "bush" back into agricultural areas which have had to be abandoned due to profound land degradation.

In England, the source of many of the plants (and animals) jammed "willy nilly" into the Australian landscape, thousands of kilometers of hedgerows full of native plants and trees were removed in the 1980's. They were considered an impediment to efficient land use. It was only later realized the protection the hedgerows gave in rainy England to the local water resources, wildlife, and soil was far more important than anyone had believed. Their removal led to serious problems including soil erosion, gully formation, and a catastrophic decline in British songbird populations.

Our Historic Relationship with Trees

Our relative cluelessness about the value of the tree in the landscape is apparently a recent development. In the Bible the righteous person was said to flourish like the date palm. Jonah, the prophet, sulking in the desert outside of Nineveh was comforted by the growth of a wild vine over his bleak brush hut and devastated when the vine sickened and died—the visceral reaction of a man living in a treeless and shadeless desert who was starved for the sight of something green.

The book of Jonah

In Native American legends, trees had characters and souls and spoke wisdom to humanity. The ancient Greeks believed trees were the homes of benign feminine spirits. Peace was made between warring tribes under the sacred indaba tree in southern Africa, a neutral place where enemies could meet. The Biblical Patriarch Abraham planted tamarisks in Be'er Sheva (a name which means "seven wells) in an expression joy and thankfulness at finding good grazing for his flocks

A man who knew the value of trees—Abraham, as depicted in the Dore Bible

©iStockphoto.com/Duncan Walker

Village life around the world was seen to flourish under various trees such as oak, chestnuts, and banyans.

Village in Bavaria—and trees
©iStockphoto.com/Richard Schmidt-Zuper

Are There Trees in Our Future?

Trees are present, then, in our histories and in our legends. So where are the trees in various "save-the-world" strategies? In the modern world, there are no sacred trees.

Global warming advocates have no problem with recommending all roofs and streets should be painted white to reflect heat back into space. They see no reason not to tamper with the upper atmosphere by filling it with reflective particles or giant mirrors. Disappointingly, only a few geo-engineers advocate tree planting programs. When they do, they concentrate on wild schemes to make artificial rivers of desalinated water run backwards into the Sahara to create a vast green belt to cool the planet—and be the lungs of the world. Of course we have such a place already—it is called the Amazon basin—and there humanity, geo-engineers included, is quickly cutting the trees down!

Where are the trees in long term planning? With all that is known about the nexus of interactions between the tree and the soil, the tree and the atmosphere, the tree and the weather, the tree and wildlife, how is it their great value and their stupendous influence can be so easily ignored? Is a civilization capable of leveling mountains, digging tunnels under the sea for high speed trains and putting people into space, incapable of planting a few trees to protect watersheds, mitigate the climate and reclaim eroded lands?

What then are trees for modern, commercial and industrial man?

> Paper and lumber.
> Obstacles to development.
> Scenery.

Tree lined bicycle path
©iStockphoto.com/Marco Maccarini

If there was any residual respect for the trees—or any vestige of the awe our ancestors once felt—then the redwoods of California would not be turned into lumber for the back deck. The fantastically tall Mountain Ash of New Zealand, the world's tallest hardwood, would not be cut down and made into disposable chopsticks for the Japanese market.

Perhaps there are few trees in our environments because there are so few in our thoughts. Before we can plant trees in the real world, responsibly use them and cherish them, make peace under them, and put them back on our farms, in our cities, and in our stories, we have to plant them back in the landscapes of our minds.

Village in India-and trees
©iStockphoto.com/Steven Miric

Chapter 16
Microstock Trees

Microstock Trees are trees used as growth habitats for creatures which are collected and harvested. These trees may be planted deliberately or they may be naturally occurring wild trees. The creatures collected and harvested include edible insects, silk producing insects, and edible mollusks.

Edible Insects

Insects have played an important role in the history of human nutrition in Africa, Australia, Asia, and the Americas. Hundreds of species have been used as human food. Some of the more important groups include grasshoppers; caterpillars; beetle grubs and (sometimes) adults; winged termites (some of which are very large in the tropics); bee, wasp, and ant brood (larvae and pupae), as well as winged ants; cicadas; crickets; and a variety of aquatic insects. Sometimes insects are used as emergency food to ward off starvation, but often they are included as a normal part of the diet either throughout the year or when seasonally available. Sometimes they are very important sources of protein in a local diet.

In Europe, the use of insects of foods has been very limited. Although frequently mentioned in ancient Greek and Roman literature, there are only very few reports on the use of insects as food in later centuries. Only in times of starvation were insects eaten among the general population. This is in stark contrast to Asia and Africa where insects are frequently considered delicacies.

Recently the use of insects as food has declined in many tropical regions, partly to increased availability of "better" foods. This often includes meat and more Western- style dishes. As insects are a very good source of nutrients, the question remains whether insects are not actually the better food from the point of view of a nutritionalist.

In Africa insects are have gone beyond being a local food and are traded on a large scale. Industries produce canned insects, dried insects, candied insects, salted insects, and insect dishes for export.

Oddly enough in the US and to a lesser extent in Europe, eating insects has increased. Not as a regular food, but more as a curiosity. For example insects are covered in chocolate or offered as sugared or crystallized candies.

Most religions accept insects as normal food and place no restrictions or taboos on consumption of insects. Jewish traditions consider only a few types of insects as kosher, almost all of them in the grasshopper/locust families. However, in practice, Jews avoid eating insects deliberately, as only trained entomologists may be able to distinguish between kosher and non-kosher insects. The exception to this rule was the Jews of the Middle East and North Africa who regularly ate locusts, crickets, and grasshoppers in times when these creatures were abundant and damaged the crops.

In Muslim regions the use of insects is very restricted. Only grasshoppers are considered halal (permissible to eat) when they have died after swarming or have been killed for food. Practically all other insects are considered unfit for food. However, in countries such as India, Indonesia, and Malaysia, many different insects are eaten traditionally, even in nominal Islamic regions. In Arab countries only grasshoppers can be found on markets.

Over 1500 different species of insects have been reported as being consumed or edible, but the insects with which this chapter is concerned are insects and other creatures which can be raised as microstock, with trees as their pastures/habitats.

Butterflies and Moths (Order Lepidoptera)

The larvae (caterpillars) of many species of moths (and a few species of butterflies) are used as food. Adult moths and butterflies are not eaten – their wings and bodies are clothed with small flat scales and hairs making them unpalatable. In Africa, larvae are a particularly important source of nutrition—protein, fat, vitamins, and minerals. More than 30 species of larvae are harvested in one African country alone, the Congo (formerly Zaire). Some caterpillars are sold not only in the local village markets, but are shipped on a large scale from one country to another. Caterpillars are canned in Botswana and South Africa. In the rural countryside, they are usually dried in the sun before being sold in the market. While many caterpillars are wildcrafted, some are being domesticated to insure a steady supply.

Caterpillars start life as eggs laid on a particular plant or tree. It is not difficult to find the eggs of the Ngala species because they are laid in a pile, like a tiny pyramid of white balls, on the new leaves at the end of a Kigala (*Crossopteryx febrifuga*) tree branch in December. This makes it easy to collect eggs and place them on appropriate plants for development. However some species, such as the Kaba worm (*Lobobunaea phaedusa*), lay their eggs singly so it is much harder to find them. Caterpillars which hide their eggs will usually hatch out undiscovered and to spread them out on tree "pastures," the tiny young have to be moved carefully to a new plant.

Young of an edible species a few days after hatching

After the eggs hatch out and the little caterpillars start feeding, they grow very quickly. Most molt or shed their old skins many times as they increase in size. Sometimes they change their color and markings, like the Makedi kedi caterpillar (*Bunaea alcinoe*), which changes from brown to red and finally to red-black in its last stage before pupation. Certain caterpillars move down the trunk of the tree on which they have been feeding in order to molt, as is the case with Ibrasia worm (*Imbrasia ertli*). It is at this time, when they have reached their full size but before they pupate, they are usually collected for eating.

Caterpillars will not appear the following year unless some have been allowed to live and pupate into adults who mate and lay the eggs which will hatch in the following year. It is normal for many caterpillars to be eaten by birds. Some will be killed by ants or by parasites. It is important, therefore, to have enough left to give a good harvest for next year. At least half the caterpillars on the tree should be left to produce sufficient numbers for next season.

When the caterpillar has reached its full size, it pupates. Some caterpillars pupate in their host trees, others in the earth, or on the undersides of leaves. Certain caterpillars, such as the Nkankiti worm (*Anaphe spp.*), form a communal cocoon in the branches of a tree. Normally caterpillars should be disturbed as little as possible, especially when they commence to pupate. If there is an aggressive caterpillar eating population of birds in the area, heavily infested trees may be netted to keep the caterpillars and pupae safe. Ringing tree trunks with borax crystals may deter ants and refraining from plowing, weeding, and "burning off" will preserve the species which pupate in the earth.

For most species the cocoon remains intact until the start of the next rainy season when the adult emerges. After mating, the female finds a suitable food plant and lays her eggs before she dies. The male dies immediately following mating.

Each species of caterpillar prefers to live on certain plants. The Ngala worm is only found on Kigala trees in Bas Congo. The Imbrasia worm lives on Kimbaki (*Funtumia africana*), Kivinsu (*Petersianthus macrocarpus*), Kinzenze (*Holarrhena floribunda*) or Kingela (Ricinodendron heudelotii) trees. It is important there are enough trees of the right kind to provide food for the caterpillars, especially those which only feed on one kind of plant. Where they are not present they can often be planted from seed, or young plants can be collected from the earth beneath mature trees and transplanted to suitable areas.

Mopane tree
©iStockphoto.com/Isabella Pfenninger

Understanding the life cycle of the caterpillar and the life cycle of the associated food plants utilized by each species will help in protecting and increasing the supply of caterpillars. The Ngala worm (Cirina forda) is a caterpillar which lives in the savannah and pupates just below the soil surface. During the dry season it is common to burn the grass in order to control rodents. However, at the same time Ngala cocoons, each capable of eventually producing far more than 50 kg of caterpillars, will have been destroyed. Ibrasia worms molt in a mass on the tree trunks and also gather in this manner to pupate. If all the Ibrasia caterpillars are taken when they gather, there will be none left to produce next year's caterpillars. If host trees are cut down during harvesting or felled for timber the entire life cycle of the caterpillar in the area may become untenable. Simple conservation measures must be taught to avoid these actions, which endanger this valuable source of food.

To quote from a caterpillar domestication program in the Bas Congo:

> *"A good place for rearing caterpillars is in a forest area which already has beehives. A good strong colony of bees will deter indiscriminate collecting of caterpillars and thus help to protect both the caterpillars and the trees. In this way a permanent reserve will be created where the caterpillars can be carefully managed. If possible it is best to have an area of at least one ha for this purpose. It may be possible to reserve a special area for the beehives for the whole village and this would also be a suitable location for raising caterpillars."*
>
> *Latham, Paul, "Edible caterpillars and their food plants in Bas Congo" (1999)*

The most popular candidate for domestication is currently the mopane worm (*Gonimbrasia belina*). The Microstock Tree which is frequently the habitat of this insect is the Mopane Tree (*Colophospermum mopane*). Domesticating the tree is probably the most logical starting point in efforts to domesticate the mopane caterpillar. While other trees in the woodlands can serve as hosts, the mopane tree is the preferred tree for food (leaves) and the favorite tree for breeding. The Anomalous Mopane Moth (Imbrasia belina) pupates in the earth near Mopnae trees.

The mopane worm, is considered a delicacy by rural and, increasingly, urban populations in southern Africa and beyond. In addition, other mopane woodland products, such as wood for crafts, fuel, and fodder for animals are important resources for poor farmers and landless poor people in the region. The trade in mopane worms is now worth several million dollars every year, but mopane worm outbreaks are unpredictable and of short duration and most of the value is captured by mobile, large scale traders rather than local communities.

Mopane worm on leaf
©iStockphoto.com/Duncan Stilwell

A good domestication program would limit the amount of damage done to the trees of the mopane woodlands by caterpillar harvesters who often fell the best trees to collect the caterpillars. Heavy infestations also reduce the amount of seeds set by the host tree; sometimes by 50 or 60 percent. Spreading the caterpillars out among many trees when they hatch could allow mopane worms to gain greater size and weight as they will be in less competition for food on their tree "pasture."

A similar strategy may be the best way to domesticate the Imbrasia worm and to protect its favorite host trees. The Imbrasia worm is usually collected when it gathers in great numbers on the light-barked trunk of the *Funtumia africana*. The caterpillars descend from the foliage of the tree each time they molt. It is at this stage they are collected for eating. They can either be eaten after roasting or boiling or else can be sun dried for later use.

A colony of caterpillars on a tree trunk
©iStockphoto.com/erin vernon

Cicadas (Order Homoptera)

This order includes many insects, such as aphids and leafhoppers, which are serious agricultural pests, but only the cicadas are used widely as human food. The nymphs of some species, known as "periodical cicadas," spend up to 17 years underground where they feed on roots. After 17 years they emerge from the soil, climb up a tree trunk or fence post and molt to the adult stage. Periodical cicadas occur as "broods" which appear above ground only once every several years in any one locality. When they do appear it is often in vast numbers. This is when they are collected as food even in the United States. They are usually fried before they are eaten. Many cicadas have shorter life cycles, and some of them were collected as food by Indian tribes in what is now the western United States, where they were sometimes dried for food for the winter and occasionally pounded up in pemmican.

They are eaten regularly in many other countries, especially in Asia, and some are very large. A cicada from Malaysia has a wing span of nearly 18 cm, larger than many birds. Most cicadas are wildcrafted by people who know what kind of trees and bushes they frequent. They pluck the cicadas off the foliage after they have finished their molt and confine them in baskets until they are used.

All deciduous woody tree species of the eastern United States serve as hosts for large numbers of root-parasitizing cicada nymphs. Periodical cicadas spend their larval lives 6 to 24 inches underground, feeding on xylem fluid from rootlets and roots. They emerge every seventeenth year in the north of the US or every thirteenth year in the south of the US. Densities of cicadas underground are very great. Emergence densities of over 300 nymphs per square yard or about 1,500,000 per acre have been reported. This represents the highest reported biomass values of any naturally occurring terrestrial animal.

During the adult stage, which lasts 3 to 4 weeks, cicadas mate and females lay their eggs on the twigs of deciduous trees. When the eggs hatch, the first instar nymphs fall to the ground to begin their slow 17-yr development. The damaging effects on twigs by egg-laying adult cicadas are serious but the major effect of cicadas may be due to the feeding of nymphs on plant roots. Feeding cicada nymphs can reduce tree growth, as measured by growth rings, by as much as 30%.

Brood X, one of the North American periodical cicadas (Magicicada spp.)
©iStockphoto.com/Brent Miles

Termites (Order Isoptera)

Termites are most widely used as food in Africa. There are sometimes called "white ants" because of the color of their backs and abdomens. Termites are social insects with colonies divided into "castes" including workers, soldiers, winged adults, and a queen. The queen is many times the size of a worker and lays thousands of eggs.

Colonies of some species build huge earthen mounds from mud, bits of plant material, wood, and a glue-like material they secrete called "termitaria." These mounds may be up to 20 feet high. Periodically, the winged adults emerge in huge swarms, mate while in flight, and then start new colonies. They are highly attracted to lights, even candlelight, and this is one way they are captured for use as food.

Termites
©iStockphoto.com/Morley Read

Logs, tree trunks and wooden beams are appealing to a new colony and are sometime placed deliberately in an area to attract new colonies. They will break down the wood as they construct their new mounds. Some species will infest the wood itself, making thousands of interconnecting tunnels around brood chambers.

As food the collected termites are captured, their wings are broken off, and they are fried and salted. Termites have a high fat and mineral content. Many people find termites delicious. The queens are considered a special treat and are often reserved for children or grandparents because they are thought to be especially nutritious and are many times the size of the common workers.

In general termites prefer hard woods, fruit woods and soft woods with little tarriness. However, termites will infest and utilize any source of cellulose, including books, newspapers, and piles of straw and sawdust.

Bees, Ants and Wasps (Order Hymenoptera)

With bees and wasps, it is usually the bee or wasp "brood" (larvae/pupae) which is eaten. Most adult bees and wasps are not palatable, but there are exceptions. In China, adult bees are fried and dusted with chile powder before being eaten. Canned wasps, wings and all, are sold in Japan, and rice cooked with these wasps was a favorite dish of the late Emperor Hirohito. In some cultures, bee nests are collected as much for their larva as for the honey. Uncapped honey combs full of bee larva are considered a delicacy.

With ants, it is also the larvae and/or pupae which are usually eaten, but adults are also sometimes on the menu. Roasted leafcutter ant abdomens are sold, instead of popcorn, in movie theaters in some places In South America. Whole ants are sometimes preserved in honey or covered in chocolate. In Mexico, certain kinds of ant pupae, known as escamoles, are found on the menu in the finest restaurants. They are served fried with butter, or fried with onions and garlic.

Many of the bees, wasps, and ants are wildcrafted, but all three types of insects are encouraged to build colonies under eaves, in trees, or in proximity to farm buildings. This is done by placing an appropriate skip or hollow log in an appealing place, or in the case of ants, leaving a trail of food from an established nest to an area where a nest of edible ants is desired. Swarming adults are likely to choose a site where an appropriate food source is nearby.

Beetles (Order Coleoptera)

Larvae, pupae and/or adults of many species of beetles are also used as food. Usually, people do not eat adult beetles whole; the hard parts (wings, legs, and head) are removed during preparation for cooking. The larvae (sometimes called "grubs") are soft-bodied. The exception to this rule is the extremely hard shelled rhinoceros beetle [a subfamily (Dynastinae) of beetles in the family of scarab beetles (Scarabaeidae)] whose huge grubs are a problem in lawns and tree roots in warmer areas of the world. These beetles are collected, roasted and ground into a spice like preparation in many countries in Asia. Their enormous grubs are consumed after being toasted. Many grubs are collected during cultivation of certain crops and many more are discovered and harvested when trees are transplanted. The roots of date trees are often infested with huge grubs. Around the roots of these trees and other palms are good places to look when searching for grubs and beetles. Beetles and grubs of edible types can also be found in rotting logs and under piles of leaves

Australian arlkerlatye grubs (Wichetti grubs)
©iStockphoto.com/Ron Hohenhaus

Grasshoppers, Crickets, etc. (Order Orthoptera)

Grasshoppers and crickets and their relatives have played an important role in the history of human nutrition. Locusts, sometimes the only food left after a swarm has destroyed the local vegetation, are an allowed food for many traditions which eschew insect eating, such as the Jewish culture. The most common way to eat locusts is by first frying them in oil. Roasting and sautéing are frequently used methods of cooking, after first removing the wings and legs. Seasonings such as onion, garlic, cayenne, chili peppers, or soy sauce may be added. Candied grasshoppers, known as inago, are a favorite cocktail snack in Japan. While locusts, grasshoppers, and crickets are attracted to certain plants, a hungry swarm will eat anything green. So far, crickets, locusts, and grasshoppers are almost all wildcrafted. Attempts at raising them deliberately on trees with abundant foliage have met with mixed success.

Heliculture

Another important food which is raised on trees and living plants is not an insect but a mollusk. Snail shells have been found in prehistoric middens all over the world showing our ancestors found them an easily obtainable food source.

It is thought water snails were first cultivated as food in China. It is also possible that terrestrial snail cultivation originated in China and southeast Asia, utilizing the lush trees of those areas. Heliculture, as the raising of snails is called, may have originated there, but it has been practiced all over the world. In an example of the simplest type of heliculture eggs of snails are deliberately moved to appropriate host trees and the trunks smeared with charcoal to keep the snails from descending to the ground. The snails graze on the trees in which they are "planted," and are harvested when large enough to be eaten.

Sophisticated snail cultivation in Europe is a branch of agriculture/husbandry known since Roman times. The Romans loved snails. They bought them from local wildcrafters and fattened them in snail gardens until they could be prepared as food. As the Roman Empire's area of influence increased, so did the distribution of snails and the market for snails. Snails shells often are found in ancient kitchen waste heaps which have been excavated around former Roman settlements.

Other peoples used snails as food and fed them on grass and agricultural by- products. The common brown garden snail was already cultivated in Celtic times. Greeks, Phoenicians, and other pre-Roman cultures in the Mediterranean region, consumed snails and other mollusks as a regular part of their diet. In the Middle Ages, snails had the crucial advantage of being neither fish nor meat, so they could be eaten during the time of Lent. Consequently most monasteries also had a snail garden, where the monks could keep snails and fatten them on fruit tree leaves, fresh grass, and bread crumbs. At that time, though, the monks were not the only people to eat snails. Snails were food for the poor too, as they could be gathered in wild areas and they were very nutritious.

Farm snail (Helix pomatia) in garlic sauce
©iStockphoto.com/ShyMan

For home consumption, it was sufficient to collect wild snails, keep them in snail gardens to increase their weight and then eat them. However, for a successful trade, snails had to be very large and of good quality. Soon it became customary to feed the snails with a diet of special foods which gave them good size and taste.

Chinese snails were fed with ground bean meal and special herbs. In Europe, snails were fattened on cornmeal and stale bread. They were transported by special boats to the big cities. There the snails were sold in the markets. There was vigorous international trade with Vienna until well into the 18th century. Later the main focus of snail trade moved towards Paris, where snails could be transported overland. In 1908 alone, the village of Guttenstein sold 4 million live snails to Paris.

Enclosure in a snail farm
©iStockphoto.com/fotolinchen

Even today snail cultivation is important. In France 40,000 tons of snails are eaten per year. A large part of those are snails are wildcrafted in natural areas in Eastern Europe and Turkey. From an ecological point of view, this is unwise and many snails are of poor quality and sometimes contaminated and may be dangerous to the consumer.

French snail cultivation methods usually aim at fattening the snails by keeping large numbers of them in small spaces, greenhouses, and pens, and feeding them with various food mixes. In France, usually the Escargot Petit Gris (*Cornu aspersum*) is cultivated. Though this species cannot compete with the Escargot de Bourgogne (*Helix pomatia*) in taste and size, it is difficult to cultivate *Helix pomatia*. Only *Cornu aspersum*, also called the common snail, due to its abundant distribution on the British Isles, is easy to raise and so can be kept in a less specialized format. A solution is sought by cross-breeding with giant *Cornu aspersum* varieties from North Africa for a bigger snail which more resembles the *Helix pomatia*.

The French still remain the main producers and consumers of snails.

Successful modern snail marketing has spread also into other countries. Snails are now raised in Germany, Switzerland, and Austria in modern formats. While cultivation habits have changed, in 18th century Austria snail farming was quite widespread. Many noble's estates had their own snail farms where the snails were raised in pens among the gardens and orchards. The new methods of snail cultivation are close to nature and reprise successful historical methods from Southern Germany, while adding modern cultivation practices. Legumes are used in crop rotation to supply natural nitrogen fertilizer Green manuring is standard practice and fresh, green food is fed instead of artificial food mixes.

A farm snail in lettuce
©iStockphoto.com/sean davis

The snails are basically kept in pens bordered by fences, in which green food plants are grown before the snails are placed there. A metal wall dug deeply into the soil around the farm keeps snail predators, such as mice, shrews and other snail hunters, out. A specially manufactured net fence keeps the snails in the enclosure.

Mangold is a favorite food of snails

©iStockphoto.com/Kurt Hahn

A special limiting factor is the natural regulation of fertility among snails which prevents overpopulation. Only 20 snails can be kept on one square meter (3300 snails are kept in an enclosure of 150 to 160 square meters). A snail's slime contains a chemical agent limiting fertility. It is disagreeable for a snail to have to crawl over another snail's trace and if they are exposed to a number of snail trails, the snails will not reproduce. Enclosures must never be too crowded and in a successful snail farm there must be several gardens to raise the young snails.

The natural prerequisite for keeping snails in enclosures is a mildly alkaline type of soil with a sufficient content of calcium carbonate. A snail farm can be sited in most places where snails survive in nature. Moderate humidity because of dew and natural precipitation is best, though snails have been raised under rainbirds and other mist irrigators. In areas with dew, the snails are usually not netted against birds, but nets can be used efficiently in irrigated snail gardens.

A sufficient amount of live vegetation in the enclosure not only provides food, but also hiding places for the snails. Food plants, such as clover, chicory, turnip, rapeseed, spinach, beets, and wild cabbage are suitable for modern snail farming. In modern snail culture very few tree leaves are used with the exception of the mulberry tree, which all grazing and browsing creatures seem to favor. A diet made up of solely fresh green food is not sufficient for good weight gain and growth. At regular intervals the snails must also be provided with dry plant food and meal.

When a percentage of the snails are moved into a new enclosure to prevent overpopulation, another part is selected to be processed. This means the snails have first to be collected, and then to be killed and sorted. If snails are processed properly, they will not suffer unnecessarily. Snails are killed in boiling hot water, as are some creatures raised for seafood. The visceral sac with digestive gland and most of the digestive apparatus is removed. This means the snails do not have to be starved before they are killed as many farm animals are.

Mobile snails - ready for marketing
©iStockphoto.com/syagci

Usually snails are preserved in different ways, such as in tins, in a sauce, or frozen. Transporting live snails is complicated, as the route of transport must remain short, and special transport containers are necessary which keep their temperatures stable, allow for ventilation, but do not let the snails escape.

Mainly two types of snails are marketed: The mobile snails collected in early summer after laying their eggs, and the fat "lid" snails (snails which have sealed themselves in their shells) The lid snails are collected after the start of hibernation, when they have the largest weight and are richest in nutritive content.

Snails at an outdoor market
©iStockphoto.com/Alison Stieglitz

Cultivating snails for food remains an intriguing microstock possibility. Cuisines outside of France have begun to show interest in snail recipes. Snail farms are also appearing in Africa where local populations are often deficient in protein. There, it is hoped the snails will be an integral part of the diet rather than a delicacy.

In Nigeria giant land snails are fed on maize chaff and paw-paw leaves. In Ghana snail enclosures are planted deliberately with selected host trees and herbs. These are only two of the dozen African countries seriously considering heliculture as a way of providing more protein to rural populations and introducing what may be a possible export crop.

Silkworms

Perhaps the most important insect raised on trees is the domesticated silkworm which is fed exclusively on the leaves of the mulberry tree and raised for the thread in its cocoon. "Silkworm" is the common name for the larva of various species of moths, indigenous to Asia and Africa, but now domesticated and raised for silk production throughout most of the temperate zone. The culture of silkworms is called sericulture.

Silkworms eat a mulberry leaf
©iStockphoto.com/Huiping Zhu

The various species of silkworms raised today are distinguished by the quality of the silk they produce, the type of leaves on which they feed, and how many times they breed within a year. The most widely raised type, and the producer of the finest silk, is the larva of *Bombyx mori*, of Asian origin. After centuries of domestication, *Bombyx mori* is no longer found anywhere in a natural state. The legs of the larvae have degenerated, and the adults are not capable of flight.

Silkworm cocoons in a factory in Vietnam
©iStockphoto.com/Anka Kaczmarzyk

Hatched from eggs so small 30,000 of them can fit in a matchbox and weigh only an ounce, these silkworms are immediately quite active on hatching and feed voraciously on mulberry leaves. At the end of the larval stage, 30 to 40 days after hatching, the worms are about 3 in. (7.5 cm) long. A full size larva attaches itself to a twig and, with a weaving motion of its head and a slow, circular motion of its body, begins to spin its cocoon. A moist substance, fibroin, is manufactured in two silk glands located on the underside of the larva's body. Mixed with a small amount of wax, it is emitted from an orifice called the spinneret, in the lip of the larva. The fibroin dries quickly in the air, hardening into a half-mile-long thread of silk which makes up the cocoon.

The adult moth, with a wingspread of 1.75 in. (4.5 cm), emerges from the cocoon in about two weeks. The moths mate and lay their eggs (several hundred from each female) within a week. The eggs hatch in about ten days. Only enough cocoons to ensure adequate reproduction are allowed to hatch. The rest are unwound after developing for a week, and the silk is processed.

The giant silkworms used in some Asian and South American sericulture are the larvae of the closely related saturnid moths (family Saturniidae). They include the tasar silkworm, mentioned below.

Giant silk moth

The ailanthus moth (*Samia walkeri*) is also a close relative. It is a large, olive-green saturnid moth used in China to produce a coarse grade of silk. It was imported into the United States along with its food plant, the Chinese ailanthus tree, as the basis of a new North American branch of the silk industry. The industry never got a start, but the moth has been firmly established in the New York City area since 1861, and so has its host tree, which first naturalized itself quite successfully in Brooklyn.

Tasar Silk Worm

The cultivated tasar silkworm is raised in China for its tawny colored silk. It is referred to as Tussah, Chinese Tussah, Oak Tussah, or Temperate Tussah. It is the source of Tussah spinning fiber which is used in the West. It is a relative of the Tropical Tussah silkmoth, *Antheraea mylitta* of India, and also related to *Antheraea polyphemus*, the American Polyphemus silkmoth.

In China, they are fed on plantations of specially trimmed oak trees on the hillsides and tended much as domesticated silkworms are pampered with optimal feeding and protection from predators and the elements.

While domesticated tasar silkworms are in no danger, the wild tasars are becoming rather rare. The wild tasars feed primarily on *Shorea robusta* (Sal), *Terminalia arjuna* (Arjun), *Terminalia tomentosa* (Asan) and *Terminalia catapp* (Sea Almond) besides a variety of secondary and tertiary food plants available in dense tropical deciduous forests of Asia. Some wild tasar species are important for the livelihood of thousands of extended Indian tribal families on the subcontinent. However, the extensive collection of wildcrafted cocoons, rapid deforestation, and human encroachment to insect habitats has greatly reduced wild populations. There is an imperative need to involve and educate local tribal peoples on sustainable cocoon collection and habitat conservation to save the wild tasar from extinction.

Tussah silk fiber dyed purple
©iStockphoto.com/Teresa Levite

This chapter concentrates on insects and mollusks which have been more or less domesticated and graze on trees the way sheep graze on pasture. However, there are many things which "grow on trees" some tame and some wild. Some are animal and some vegetable. Some are used for food and some for medicine. The trees which support these organisms are commonly called Host Trees. They are the subject of the next chapter.

Chapter 17
Host Trees

From an agricultural viewpoint the most interesting organisms which use trees as hosts are fungi. "Fungi" is the name given to group of living things including mushrooms, toadstools, moulds, truffles, slime, and by stretching the definition a bit, lichen (lichen is a symbiosis of fungi and algae). Unlike green plants, Fungi cannot produce their own food. Instead they absorb nutrients from their surroundings.

The fungi which grow on live or dead trees are highly specialized. There are saprophytes, which live on dead stumps and logs, and also symbiotic fungi which grow on or in living hosts.

Symbiotic Fungi

Although these fungi do live on or inside other living things, they do not cause damage. The fungi, and the organism on which it lives, both receive benefits from living with each other. Many types of fungi live in such harmonious "give and take" relationships with trees, and other plants, including orchids. The fungi grow underground, and their threadlike "hyphae" grow into a thick mat known as "mycelium." This mycelium absorbs more nutrients and water from the soil than it needs and passes the excess to the tree through the roots of the tree. Such relationships between fungi and trees are known as mycorrhizal relationships. These fungi are more or less invisible, unless one is looking for them, and of course are not harvested or used for food. However, whole other classes and families of fungi use trees as hosts including truffles, mushrooms, bracket fungi, and puffballs. Many of them are highly prized as delicacies or medicinal material.

Some have already been domesticated. As an example of a domesticated fungus which needs living trees as a host, one need look no further than the truffle. Truffles are a hypogenous fungus, which means they form their fruitbodies below ground. Ecologically, they are mycorrhizal, forming mutually beneficial associations with the roots of plants. Taxonomically, they are members of the Ascomycota phylum of the fungi kingdom (having a saclike ascus which contains the ascospores).

Like most other hypogenous fungi, wild truffles depend on animals to disperse their spores and help them reproduce. The ripening truffles emit a distinct aroma, which grows stronger as they mature. The aroma attracts a variety of animals (humans included) who eagerly collect and consume the truffles, and later disperse the spores to new areas, by contact with the truffle, their feces, or when speaking of humanity, by deliberate inoculation to new host plants.

Truffle Cultivation

Truffles are sold for between 800 dollars a pound and 3,000 dollars a pound. A successful truffle orchard can produce 130 to 150 pounds per acre per year on land which is good enough to grow a selection of hardy forest trees such as oaks, elms and hazelnuts. The area chosen for a truffle orchard must also fulfill other soil and climatic conditions.

The combination of high prices and the availability of areas in the temperate zones which may be suitable for cultivation of the host trees has made truffle cultivation an extremely promising new branch of agricultural endeavor.

Area Must Have High pH Soils: The ground for the truffière (truffle orchard) must be high in pH. A minimum pH level is 7.5 and the optimum is 7.9. Many of areas otherwise suitable are well below this level and if a farmer seeks to establish a truffière on low pH soil, he/she must make the effort to lift the pH and keep it permanently high. The best course is to find a part of the property where pH is naturally high or to move to a location where the soil pH is naturally high.

Area Must Have Free Drainage: The development of the truffle is inhibited in clay soils. If the soil becomes too wet the truffle can also rot underground before it is ready for harvest. In order to develop the truffle bed the top 300mm of soil should be loose and friable enough to enable the farmer to dig out the truffles when they develop.

Area Must Have Good Quality Water For Irrigation: Irrigation water must be available for the site when required. Although the truffière need not be irrigated all the time, the fungus develops poorly if the ground dries out for extended periods. Supplementary irrigation may therefore be required and the soil humidity should be measured regularly. The guiding rule with irrigation is, if the trees are suffering, the truffles will suffer as well. The water source must not only be plentiful and reliable, it must be of high quality.

Area Must Not Have Chemical Residues In The Soil: Truffles are very sensitive to chemical residues in the soil. Site selection for the truffière should take into account the previous land use and stay away from sites or cropping areas which may contain herbicide or pesticide residues.

Host Trees: The most common host trees planted in truffière are European: English Oak (*Quuercus*

Robur), Holly Oak (*Quercus ilex*) and Common Hazel (*Corylus avellana*) but birch, alder, elder, and elm trees are also used. To prepare saplings as host trees, they are inoculated in the nursery ensure the growth of the fungi. The Oak and Hazel trees create the best conditions for the black Périgord truffle (*Tuber melanosporum*).

Wind Protection: Small trees need some form of wind protection when they are still young. However, this is not necessarily achieved by the planting of native tree belts because these can contaminate the truffle bed with wild truffle species.

Fencing: In the early stages of truffle bed development, the young trees are vulnerable to rabbits and other wild life. A small meshed fence too high to jump, on sturdy posts is required to keep out all small grazing animals. To keep out rabbits and other gnawers, it is best to set the fence in the ground by at least 20 cm.

Buffer zone: Due to the potential for wild fungal contamination it is best to create a border-zone around the truffle bed. This may be in the range of 50m - 100m.

Climate: A climate with crisp frosty winter mornings and plentiful sunshine hours is ideal. Most truffle trees are temperate zone deciduous trees. Truffles will grow where summer temperatures are hot and dry, but they will need to be irrigated if cultivated in such an area.

Who finds the truffles? Truffles can sometimes be spotted by areas of dead grass that their ripening produces. These are called “brules.” A prototype electronic truffle sniffer is under development to help find the truffles, but the most effective method of truffle detection is by using the superior sense of smell of some animals. Traditionally, in France, pigs and dogs have been used for this purpose. Pigs are not as easy to train and restrain as dogs, so currently dogs are more popular among truffle farmers. Once located, the task still remains for the farmer to dig carefully with a pronged fork to recover the truffle without damaging it. This is why the upper layers of the earth in the truffle orchard should be loose and friable.

Freshly harvested truffles
©iStockphoto.com/Alain Couillaud

When to harvest: The complicated root matting and the development of the truffle fruiting bodies is a lengthy process. First harvest in a truffière may be from four to seven years after planting.

Yield: The yield of a truffière is measured in kilograms of truffles per tree. Once the trees are mature, each tree has the potential to produce up to 1 kg of truffles per year. This is a reasonable planning figure. There have been cases of trees yielding up to 2 kg per year.

Growing/Harvest cycle: Most truffle trees are deciduous. They emerge from their dormancy period in spring and grow strongly through the summer. The truffles are most active at the same time the trees are most active. In winter, the trees become dormant and so do the truffles. It is in the winter dormancy the truffles form into the hard, black, knotted tubers beneath the ground.

Truffles are among the world's most sought after culinary delicacies. Prized for their taste and texture for millennia, their price has risen until truffles have become the most expensive foodstuffs. Retail prices in the U.S. for the French black truffle or black Périgord truffle (*Tuber melanosporum*), and the Italian white truffle (*Tuber magnatum*), have reached $1,000 and $3,000 per pound. Most truffles are harvested in the wild, but since the wild supply is diminishing, prices continue to climb. This is one of the many incentives for the domestication and cultivation of these unusual food plants.

The breakthrough which made their cultivation possible was development of technology to properly inoculate host trees with the fungus under exact and controlled conditions while the trees are still in the nursery. The use of inoculated trees to cultivate *Tuber melanosporum* and other truffles has proven successful over the past two decades in Europe and many farms, including several in the United States and one in Canada, are now producing French black truffles for the market.

To cultivate truffles, inoculated truffle trees are planted in orchard formats much like those for fruits and nuts. The trees are inoculated as saplings and then planted out in optimal conditions of spacing and water availability. Truffles begin to appear several years (4 to 7 years) after the inoculated seedlings are planted out. Truffle production, once established, can continue for decades. The onset and duration of production depends to some extent on the interplay of truffle species with the host trees. Some trees are slower to take the inoculation and some truffles are slower to develop.

Yields vary dramatically from farm to farm: some farms produce as much as 150 pounds per acre each year while others produce only a dozen kilograms. Typical yields in Europe range between 25 and 35 pounds per acre each year in orchards based mostly on inoculating wild trees. As methods improve, many more farms are achieving yields in excess of 100 pounds per acre, an achievement which makes truffle cultivation a financially viable form of agriculture.

The famous black truffle *(Tuber melanosporum)* grows beneath relatively isolated trees, meadow trees, or trees at the edge of forests in its natural habitat. Thus, many plantations in Europe are sparsely planted with as few as 100 or fewer large trees per acre. In other cases, the trees are packed close with as many as 1,000 trees per acre to encourage the inoculation of all the trees by the movement of the fungus from one tree to the next through vegetative growth of the fungal mycelium. The spacing of the trees also depends to a large extent on the size of the tree species used, soil fertility, and the willingness of the farmer to thin the trees when population density exceeds healthy level, possibly removing some trees which may be producing truffles.

In North America host tree seedlings such as oaks, hazelnuts, and elms are inoculated with *Tuber melanosporum* and other truffle species. These seedlings are checked individually to confirm the presence of the proper fungus and to confirm they are abundantly colonized by the truffle ectomycorrhizae and completely free of any competitor species. The inoculated trees are then planted out in areas with the proper soil fertility, amount of organic matter and proper ph.

The approaches necessary to maintain these conditions can vary from place to place and different farmers may want to use different methods depending on availability of equipment, time, and money. The basic management practices include irrigation, weed control, soil aeration, pruning, thinning, mulching, and in some cases, fertilization. A fundamental requirement in all cases is to maintain the soil pH necessary for truffle production.

At the extremes of low and high intensity management are the Tanguy and Pallier methods of cultivation representing two distinctly different cultivation methods.

At the low intensity end of the scale the Tanguy method calls for mowing to control weeds, but does not involve soil aeration, pruning, irrigation, or fertilization. This method is simple enough for those farmers who do not own tractors, nor have abundant availability of irrigation water nor the time and money to invest in more intensive management approaches. It is also safer in the sense it errs on the side of non-interference and caution towards the development of the truffles and the host species. Because of this tendency to err on the side of less interference and intervention, there is less potential to damage the plantation. However, it generally takes a couple more years for truffle production to begin.

The more intensive Pallier method calls for light tilling or harrowing in the spring and early summer to control weeds and aerate the soil. The trees are shaped to maximize penetration of sunlight through the canopy and warm the soil. And finally, irrigation is supplied as necessary to emulate the summer and fall weather which produces the largest truffle yields. It is possible to till too deeply and damage roots, and to irrigate too much, giving the competitive advantage to other water loving fungi. These methods should be used carefully. However, the Pallier approach is thought to produce truffles somewhat earlier than less intensive approaches and to allow the farmer more control over microclimatic conditions. In some cases, irrigation may be necessary simply to keep the trees and the truffles alive if natural precipitation is insufficient.

The first cultivated truffles may be harvestable five years after the inoculated trees are planted out. They are generally detected by the presence of dried and burned looking patches of grass around or near the trunk of the tree (brules). These brules are formed when the truffle begins to mature and give off chemicals substances which kill the grass. It is at this time, the truffle begins to have the odor which makes it detectable.

An antique illustration of a truffle hunter with trained pig
©iStockphoto.com/nicoolay

In some truffle orchards a trained dog or a trained pig will indicate where to dig. The digging process is actually done by the truffle farmer (since if the animal finds the truffle he will eat or damage it) and the animal rewarded by a slice of truffle or some other treat.

Truffles in the Negev (and their hosts)

Truffles that resemble potatoes
©iStockphoto.com/anzeletti

"Desert truffle" is a term used to refer to members of the genera Terfezia and Tirmania in the family Terfeziaceae, order Pezizales, which grow in arid and semi-arid areas of the Mediterranean region, the Arabian Peninsula, and North Africa. Some desert truffle species have been found in South Africa and China. Dr. Nissan Binyamini is the local expert on the Negev truffle. Dr. Binyamini wrote the only mushroom guide for amateurs in Hebrew, and headed the Dept. of Botany at the Tel Aviv University for many years.

Species of Terfezia and Tirmania prefer high pH calcareous soils. High pH and the presence of calcium carbonate are typical of desert soils.

Although the genera Terfezia and Tirmania are primarily ectomycorrhizal (they usually form a sheath around the roots of their host plant rather then growing from the root tissue), they are highly adaptable. Some species, like *Terfezia arenaria*, *Terfezia claveryi*, and *Tirmania pinoyi*, form endomycorrhizal associations in phosphate-poor soils and ectomycorrhizal associations in phosphate-rich soils.

Species of both genera form mycorrhizas on roots mainly of members of the genus *Helianthemum* (family Cistaceae), relatives of the North American rock rose, but can also form relationships with members of other families in the absence of species of *Helianthemum* including many desert trees. These relationships contribute to *Helianthemum's* adaptability to drought conditions and facilitate absorption of nutrients, particularly nitrates.

This finding may provide an explanation for folklore shared by Bedouins in the Israeli Negev, Israeli wildcrafters, and truffle hunters in Morocco, claiming truffles will grow where lightning strikes during thunderstorms. (Lightning of course, is the original nitrogen fixing process, turning the nitrogen in the air into a form which can be used by plants.)

Plenty of rain in the beginning of the rainy season is necessary to ensure a good truffle crop in spring. Even so, many truffle hunters, including the Bedouins of the Negev, used to believe truffles appear suddenly, without seed or root, swollen by early season's rains, and loosened from their sandy bed by the loud rumblings of strong thunderstorms. These beliefs go back thousands of years and were shared by many peoples.

In the 1st century C.E., Pliny the Elder (Gaius Plinius Secundus), wrote in his *Naturalis Historia*, Book xix:

"Among the most wonderful of all things is the fact that anything can spring up and live without a root. These are called truffles (tubera). They are surrounded on all sides by earth, and supported by no fibers. There are two kinds: one is sandy and injures the teeth, the other without any foreign matter. Those of Africa are the most esteemed. Peculiar beliefs are held, for they say they are produced during autumn rains, and thunderstorms especially, and are best for food in the spring. They grow…where there is much sand."

The Jewish Talmud (the record of rabbinic discussion and extrapolations of the Jewish Torah), echoed the same claim. Truffles and mushrooms are usually discussed together in the Babylonian Talmud (compiled and redacted in Iraq in the 5th Century C.E.). The Rabbis considering the issue concluded truffles and mushrooms do not grow from the soil. Rather, they spontaneously appear in the soil. In one place it is said, "They emerge as they are in one night, wide and round like rounded cakes." This was considered a great wonder and there was some discussion among the Jews of the day as to which blessing should be employed before eating them since the ones appropriate to ordinary foods might not apply.

The most common species of the genus Terfezia are Terfezia arenaria (syn. T. leonis), T. boudieri, T. claveryi, T. leptoderma, and T. terfezioides (=Mattirolomyces terfezioides) and although just separated - T. pfeilii (syn. Kalaharituber pfeilii). The most common species of the genus Tirmania are Tirmania nivea and T. pinoyi (syn. T. africana). Terfezia spp. have spherical and ornamented spores, while Tirmania spp. have smooth spores.

The truffles themselves are round, tan to brown, and look like small, sandy potatoes. They are a few centimeters across, and weigh about 1 to 10 oz. The truffles produced by both genera are similar in overall appearance and are not easy to tell apart. Their unique flavor develops as they reach maturity and they are generally cooked in the most simple manner often by roasting on a bed of coals.

These truffles are generally found by careful examination of the soils in which they and their host plants grow. Desert truffle hunters rely on a variety of signs, including the presence of dried hyphae to lead them to the truffles.

Although the field of trufficulture has greatly expanded since its inception in 1808, several species still remain uncultivated. The desert truffle is one of them, although there is some research going on in Israel towards its domestication. There is also at least one Israeli project which is attempting to adapt the European truffle to desert cultivation conditions.

Morel Mushrooms

Beautiful, tasty, exotic, and strange, the morels are the choice tree-associated fungi of the spring.

Typically they are found in moist areas, around dying or dead Elm trees, Sycamore and Ash trees, old apple orchards, and in mixed deciduous forests. The areas where morels are found vary. It is very likely each patch of mushrooms may be growing in slightly different conditions. Morels tend to grow in or very near places in which they were found the year before, though. Wildcrafters of morels carefully mark and map these spots.

Morels are not universally appreciated. There is a certain amount of risk in eating them. One of the reasons morels are suspect is the presence of false morels in many areas, morel mushroom look-alikes which are not edible. In part, it is because of these potentially dangerous mushrooms a drive to domesticate the morel has begun.

One man seems to be well ahead of the pack in morel domestication, Stewart Miller of Indiana, a former biology teacher, who for most of his life has been very interested in mushrooms and edible fungi of all kinds. Miller began his morel mushroom cultivation research in 1972 while teaching Biology at Marion High School in Marion, Indiana. He has been studying the relationship between morels - also known as sponge mushrooms - and elm, apple, and ash trees since 1992. After years of experimentation, he was awarded US Patent Number 6,907,691 B2 in June 2005 for his morel cultivation process.

To explain how the morel mushroom forms in nature Miller coined the term "symbiotic disruption" to explain the morel mushroom's eccentric growing habits. A suffering or dying tree stimulates the morel fungus inside the root system, causing it to withdraw. Hardened nodules called "sclerotia" form below the ground, then with sufficient water and warmth in the spring, these sclerotia swell and form a morel mushroom.

Morel mushrooms
©iStockphoto.com/Becky Swora

An avid morel mushroom hunter since childhood, Miller explains growing morels eliminates the guesswork in judging whether wild morels are safe. At his Morel Farms, which consist of three different parcels of land, including a 45-acre tract near Lafayette, Indiana, Miller is growing a combined total of 2,000 apple, 3,000 ash, and 5,000 elm trees.

The morel mushroom harvest begins in April or May, depending on climate and region. In a test-plot in the fall of 2008 he planted 1,200 elm trees inoculated with his patented morel fungus, in an open, well-drained, and fertile acre of land. “In seven years, according to our projections,” said Miller, “each tree will produce approximately five morels for a total of 6,000 morels. Total income for the project would be approximately $7,500. If you divide $7,500 by seven years, the resulting income would be projected at $1,071 per acre. This and other forms of research will verify our predictions of morel mushroom production and morel farming as a viable crop.”

This and other domestication and cultivation projects may make the morel as at least as easy to obtain and cultivate as the domesticated truffle and remove much of the danger from eating morel mushrooms.

Domesticated Mushroom and Fungi

Many mushrooms and fungi have been domesticated and are cultivated like any other agricultural crop. Most of these are the type of fungi which get their energy from decaying material and so are easier to cultivate, as they can be grown on inoculated logs or other formations of cellulose which contain the proper nutrients rather than cultivated on or by host trees.

Mushroom farm
©iStockphoto.com/doga yusuf dokdok

The most popular is the White Button (*Agaricus brunnescens*) which is also the most frequently used of all mushrooms. Also, called Champignon, they have been cultivated by the French since the 1700's. Today, the United States is the largest grower of cultivated White Button Mushrooms. The Champignon retains its shape well when cooked, although they do shrink a little in size. The Champignon has a mild flavor and firm texture and comes in sliced form. Most commercially canned mushrooms are usually of this variety. The Champignon absorbs flavors well in any dish and reconstitutes well from its dried form. It has been grown on wood shavings and fine tree litter but its most common growing format uses trays of composted manure.

The first domesticated mushroom of all may have been the Chinese Wood Ear, a fungus which does grow on a living tree. The domestication and the cultivation of these mushrooms were recorded early in

Tang Dynasty times (618 - 907 CE). There is information suggesting the Wood Ear (or Cloud Ear) mushroom was the first mushroom to be domesticated in Chinese history. Before domestication, they and all other wild mushrooms were wildcrafted and often considered the food of the elite. By Sung Dynasty times (960 - 1279 CE), there was a guild of mushroom gatherers, mushroom farms, and a monograph written about wild edible mushrooms in the royal library.

The botanical name of the Wood Ear family of mushrooms is *Auricularia auricular.* They have a relative called the Silver fungus or White Tree Ear whose botanical name is *Tremella fuciformis.*

The tree ear mushroom is the only mushroom among all edible mushrooms, whose taste is the same fresh or dried (after reconstituting them by soaking). The texture is close to the same, as well. Tree or wood ears are traditional food in China. There are many varieties of mushrooms in this family, about fifteen or twenty species. They earned their name because they grow on wood and their convoluted shape resembles ears.

Tree Ears grow in bunches on the trunks of live forest tees. One species particularly prefers the elder trees and is only found on them, but most will colonize other broad leaf tree varieties. Though Oak is the most common host tree, they can be found on banyon, birch, poplar, ash, and elm trees. Some types grow on fir trees and even on pines. While these fungi are most successful when introduced to living hosts by the transfer of spawn and spores, they can be grown on harvested logs or artificial logs incorporating diverse agricultural wastes such as cotton seed shells, various types of sawdust, sugarcane, rice straw, and corn residues into an artificial matrix.

Wood Ears are believed to be healthy food and have considerable incomplete protein—many amino acids—in their composition. One hundred grams of dried Tree Ears are said to have almost eleven grams of protein, almost no fat, and sixty-five grams of carbohydrate, close to four hundred milligrams of calcium, half that amount of phosphorus, almost the same amount of iron, and various polysaccharides. It is possible they were the first mushroom to be domesticated because of their nutritive value.

Modern Domestications

The domestication of edible mushrooms and fungi has accelerated in the last decades. It is now possible to grow many types of edible fungi, whether in special formats or at home on with spawn supplied in kits. The following species are cultivated on harvested or artificial logs.

Shiitake Mushroom (Lentinula edodes)
©iStockphoto.com/merrymoonmary

Growing Temperature: 50–80° F

This mushroom is esteemed for both its health-supporting properties and its culinary value. It is grown on logs of several different forest trees including oak, alder, and birch. Shiitake mushrooms can be harvested at 2 week intervals for up to 16 weeks.

The Pearl Oyster Mushroom (Pleurotus ostreatus)
©iStockphoto.com/Roger Whiteway

Growing Temperature: 55–75°

Pearl Oyster Mushrooms start on logs but will flourish in other moist, nutrient-rich areas, like a compost pile. Often gardeners can enjoy continued harvests of Oyster Mushrooms for many months.

The Blue Oyster Mushroom (Pleurotus ostreatus var. columbinus)
©iStockphoto.com/Werner Münzker

Growing Temperature: 45–65° F

This fascinating fungus is a cold-weather variant of the Pearl Oyster Mushroom (*Pleurotus ostreatus*). It requires temperatures of approximately 65° Fahrenheit to produce mushrooms, though the mycelium will continue to grow at temperatures above 65. Due to its unique temperature requirements, the Blue Oyster mushroom can be grown in cool spaces such as basements and on the north sides of buildings.

The Flamingo Oyster (Pleurotus djamor) honored by a stamp
©iStockphoto.com/jim pruitt

Growing Temperature: 70–85° F

A tropical species, the Pink Oyster Mushroom is also known as the Flamingo Oyster.

The Nameko Mushroom (Pholiota nameko)
©iStockphoto.com/Frans Rombout

Growing Temperature: 50–65° F

The Nameko Mushroom is Japan's second most popular cultivated mushroom Shiitake is the most popular. This mushroom has a strong flavor, faintly reminiscent of cashews.

The Enokitake Mushroom (Flammulina populicola)
©iStockphoto.com/Ming Onn Boey

Growing Temperature: 45–75° F

A small capped, long stemmed mushroom a favorite in Japan and known by North Americans as The Winter Mushroom.

The Lion's Mane Mushroom (Hericium erinaceus)

Growing Temperature: 65–75° F

An extraordinary mushroom, the Lion's Mane (also called Bearded Tooth Mushroom or Hedgehog Mushroom) is one of the most unusual edible fungi. It is cold tolerant and found on hardwoods in nature, especially beech trees. It is being investigated for its influence on auto-immune and degenerative diseases. So it may be more than a tasty fungus. The fruitbody produces cascading, icicle-like clusters which can grow to the size of baseballs or larger. A choice edible, this mushroom imparts a seafood-like flavor when cooked with butter and onions and easily absorbs the flavors of spices.

The Maitake Mushroom (Grifola frondosa)
©iStockphoto.com/Yosef Galanti

Growing Temperature: 55–70° F

Maitake mushrooms are succulent and delicious. They are known in Japan as Maitake (also "Hen of the Woods" or "Sheep's Head"). They are a large, hearty mushroom commonly found on or near stumps and bases of oak trees. They are believed to have medicinal properties. They are also one of the most popular and widely recognized mushrooms in Japan.

The Reishi Mushroom (Ganoderma lucidum)

Growing Temperature: 70–80° F

Ganoderma lucidum are known as "Reishi" by the Japanese and "Ling Chi" by the Chinese. This mushroom has long been sought after for its beneficial medicinal properties. Ling Chi is perhaps the best renowned of all the traditional therapeutic fungi, represented in Asian art for thousands of years. Reputed to have many health-supporting properties, this mushroom is generally broken up, powdered, and steeped in simple teas. Its flavor is strong, distinctive, and pleasant to most people.

The Pioppino Mushroom (Agrocybe aegerita)
©iStockphoto.com/ilbusca

Growing Temperature: 50–70° F

Popular in Italy, Pioppino mushrooms have a mellow and attractive flavor. This table-top mushroom features beautiful membranous rings on the stems which slowly fall as the mushrooms mature; a study in the beauty of nature. Once fruited, The Pioppino Mushroom is an excellent candidate for stump recycling of willows, bay, alder, cottonwood, elm, and many other woods.

Mushroom kits from many companies now offer plug spawn of a number of hardy mushroom species: Reishi, The Oregon Polypore, Maitake, The Conifer Coral, Lion's Mane, Shiitake, Pearl and Phoenix Oyster, Chicken of the Woods, and Turkey Tail.

These sterilized wooden dowels are spirally grooved and fully colonized by pure mushroom mycelium.

They are sold with instruction booklets so the amateur mushroom grower will know how to grow the mushrooms, on what wood the spawn will thrive, how to harvest them, and how to cook and eat them.

Growing mushroom on logs
©iStockphoto.com/Hsing-Wen Hsu

By using the dowels to inoculate cut hardwood logs or stumps, mushroom mycelium can be encouraged to colonize the chosen wood. Once the wood is fully colonized (typically 9 to 12 months) mushrooms will grow copiously from cracks or channels in the wood. Generally, the best time of year to inoculate logs and stumps is in the spring, after the last hard frost. However, in all but the coldest of cold areas, logs can be inoculated any time up to 2 to 3 weeks before freezing temperatures set in for the winter. The idea is to allow the mushroom mycelium growing on the inoculated dowel enough time to establish itself in its new home before it goes into dormancy over the winter. Logs can be left outdoors over the winter, under a layer of straw or leaves, or a burlap tarp, shade-cloth, or other vapor-permeable cover. Do not use plastic tarps; this can cause mold to form. In areas where the winter is exceptionally harsh, logs can be stored in a shed, barn, garage, or other outbuilding.

Most inoculated spawn prefers to grow on hardwoods, with the exception of the Phoenix Oyster Mushroom, which grows well on firs. Most species can be grown on either logs or stumps. Non-aromatic hardwoods such as oak, poplar (cottonwood), elm, maple, and similar woods are very good candidates for log cultivation. Alder is a good wood for the cultivation of Oyster and Shiitake mushrooms, but must be kept above ground because it will decompose quickly in contact with the soil. Thick-barked woods are preferable over "paper-bark" woods such as birch, although birch is an excellent media for the spawn inoculated dowels. Any log which is shedding its bark should not be used. Logs should be cut one to three months in advance of inoculation. Cutting logs in the late winter or early spring helps to insure they have a high sugar content, although this is not strictly necessary. Freshly-cut logs should not be immediately inoculated. Trees naturally produce anti-fungal compounds, which degrade in two to three weeks from cutting. Aged deadwood is also not recommended for inoculation, as it has a poor nutrient base for supporting mushroom growth. Logs or stumps with fine cracks running through them are more quickly colonized with mushroom mycelium than those without.

Logs should be cut to lengths of about a meter, and are best if they do not exceed 40 cm in diameter. Holes should be drilled to allow for a snug but adequate fit for the inoculated dowels. Stumps should be inoculated along the circumference of their face, in the border between the bark and the heartwood. The more dowels used per log, the faster the wood will be colonized with mushroom mycelium. Holes can be sealed with cheese wax or beeswax to protect the mycelium from weather and insects while it is growing; although this step can be helpful, it is not absolutely necessary.

Wild Mushrooms (and their trees)

There are many wild edible fungi which grow in association with various forest trees. There are too many to list all of them, but some of the more well known edible types are found under on or under common forest trees.

Leccinum scabrum

- The *Leccinum* genus includes two well-known mushroom species named after the trees they can usually be found next to. The *Leccinum aurantiacum* (as well as the *L. versipelle*), found under aspen trees, and the *Leccinum scabrum* (as well as the *L. holopus*), found under birch trees. The two species are significantly different in cap color only. Both are very sought after, being highly palatable and beautiful.
- Armillaria - Autumn Stump-Grower; The Honey Mushroom, Shoestring Rot. The genus *Armillaria*, with the popular species *Armillaria gallica* and *A. mellea*, being so similar they are rarely differentiated, are palatable, highly abundant mushrooms. Generally found on decaying tree stumps, they grow in very large quantities and are easy to spot and identify, arguably reducing the fun and challenge in mushroom hunting.

- *Pleurotus ostreatus* - The Oyster Mushroom. It is the most commonly picked tree-dwelling mushroom and is often also artificially cultivated for sale in grocery stores. This sturdy mushroom can be quite palatable when young. Growing these mushrooms at home can be a profitable enterprise.

Picking wild forest mushrooms
©iStockphoto.com/intst

- *Tricholoma matsutake* - = syn. *T. nauseosum*, the rare red pine mushroom which has a very distinct and fine aroma. Its undeniable fragrance is both sweet and spicy. These mushrooms always grow under trees and are usually concealed under fallen leaves and the mast or duff layer. This type of mushroom forms a symbiotic relationship with the roots of a limited number of tree species. In Japan it is most commonly associated with Japanese Red Pine. However in the Pacific Northwest it is found in coniferous forests of Douglas fir, Noble fir, sugar pine, and Ponderosa pine. Further south, it is also associated with hardwoods, namely Tanoak and Madrone forests. The Pacific Northwest and other similar temperate regions along the Pacific Rim also hold great habitat producing these and other quality wild mushrooms. In 1999, N. Bergius and E. Danell reported Swedish (*Tricholoma nauseosum*) and Japanese matsutake (*T. matsutake*) are the same species. The report stimulated the import of these mushroom from Northern Europe to Japan because of the comparable flavor and taste. Matsutake are difficult to find and are therefore very expensive. Moreover, domestic productions of Matsutake in Japan have been sharply reduced over the last fifty years due to a pine nematode called *Bursaphelenchus xylophilus*, and it has influenced the price a great deal. The annual harvest of Matsutake in Japan has since further decreased. The price for Matsutake in the Japanese market is highly dependent on quality, availability, and origin. The Japanese Matsutake at the beginning of the season, which is the highest grade, can go up to $2000 per kilogram (comparable to most truffles), while the average value for imported Matsutake from China, Europe, and the United States is only about $90 per kilogram.

- The *Tricholoma magnivelare* is a very popular and commonly cultivated mushroom in North America. It has proved to be one of the easier tree associated fungi to cultivate as it can be seeded by mycelia transfer. British Columbia exports large quantities of this mushroom overseas to Asia where it is in high demand.

All of these species require association with a live tree to produce mushrooms, which makes them candidates for cultivation in forested areas not suitable for conventional agriculture. Of these so far, only chanterelles (*Cantharellus spp.*) and the Tricholomas have been grown "in captivity," and then only by achieving successful inoculation. There is some question as to whether or not the other species can be introduced into a chosen outdoor plot complete with host trees where they are not already growing, either by spore slurries or mycelia transfer. Certainly the attempt should be made as these fungi are all excellent foods.

Beware of wild mushrooms and fungi

Some mushrooms are deadly or extremely hazardous when consumed. Others, while not deadly, can nevertheless cause permanent organ damage. If using and gathering wild fungi it is strongly advised:

- Only gather mushrooms which have been positively identified.
- Do not depend on generalizations such as "all shelf and bracket fungi are edible" or "all tree mushroom are edible" because they are not true.
- To identify the mushrooms a second time during preparation, and to prepare them properly (only very few species can be eaten raw). It is considered unwise for collectors to combine the mushrooms they have collected into salads or casseroles.

- To inform oneself about deadly mushrooms and the deadly look-alikes of edible ones. The second category varies across world regions, so it is important to take into account regional variations.
- Not to gather mushrooms which are difficult to identify, unless one has an expert's knowledge. This applies especially to the mushrooms of the genus *Amanita* or *Cortinarius* and "little brown mushrooms" which are often assumed to be edible because of the appearance.
- Only to consume a small amount of the mushroom the first time one tries a certain species. People react differently to different mushrooms, and all mushroom species can cause adverse reaction in a few individuals, even the common champignon.

Integrating Fungi Growing with Conventional Agriculture

Small-scale mushroom production represents an opportunity for farmers interested in an additional enterprise and is a specialty option for farmers without much land or with forested land unsuitable for other forms of agriculture. Market gardeners who want to incorporate mushrooms into their systems and for those farmers who want to use mushroom cultivation as a way to extract value from woodlot thinning, stumps, logs, and other "waste" materials or indeed to use the living trees of the woodlot themselves as hosts should consider edible fungi as a possible crop. Mushroom production can play an important role in managing farm organic wastes when agricultural and food processing by-products are used as growing media for edible fungi. The spent substrate after harvest can then be composted and retuned to the soil. The host trees, of course, will continue to contribute to the health of the garden or farm.

Afterword

A Blessing to the Earth

I would like to conclude my book with a quote from E.F. Schumacher, the author of *Small Is Beautiful: Economics as if People Mattered*, Harper Perennial (September 27, 1989). On reading *Tree Crops: A Permanent Agriculture*, by J. Russell Smith, Shambhala Publications (December 1987), one of the most important books ever written about the subject of silviculture or arboreal agriculture, E F. Schumacher said:

> *"Tree Crops made so much sense to me that I have never been the same since. It made sense because it did not merely state that "civilized man has marched across the face of the earth and left a desert in his footprints"—a remark I found confirmed in innumerable places throughout the world; no, it did more than that: it showed what could and what should be done. Most improbably, it seemed to me, the answer has been there all the time and was still available to us. Agriculture is for the plains, while silviculture is for the hills and mountains. When the plough invades the hills and mountains it destroys the land…just as efficient agriculture depends on human ingenuity and work—in finding the best methods of cultivation, in plant breeding and so forth—so an efficient silviculture depends on the same kind of effort. Without the effort, nothing much can happen.*
>
> *"As my work took me all over the world, everywhere I could see it thanks to Russell Smith: Agriculture in rocky, dry, and mountainous regions is a disaster, but trees are salvation. And trees yielding annual crops did not have to be created, they existed already. But our care and attentions, selection and plant breeding, the application of methodical science, could improve them beyond our imagination.*

> *"All my life has been a discovery of the generosity of nature. I started out thinking we had to do everything ourselves and of course, we couldn't. But then I discovered that everything will be done for us, provided only that we realize our 'nothingness' and therefore start to search for a way of fitting in with the great processes of nature, making the most of them, for our purposes.*
>
> *"By the means of trees, wild life could be conserved, pollution decreased and the beauty of many landscapes enhanced. This is a way or at least one of the ways, to spiritual, moral, and cultural regeneration."*

I, too, read *Tree Crops* by J. Russell Smith and I, too, was changed forever. The course of my life was changed with me. The work I wanted to do, the contribution I wanted to make to those who come after me, the potential of tree crops I knew and wild trees no one seemed to know anything about—all became clear to me.

I went to live in a hard place, an unforgiving place with blazing sun, caustic soil and bitter water. There I learned tree by tree what a blessing trees are to us and to the earth.

Trees are the upwelling zones of the world of agriculture, the springs of fertility and abundance for the natural world. They will protect the earth for us, stabilize the climate for us, bring down the rain for us, store water and carbon for us, turn dry sands into fertile soil, feed, clothe, shelter, protect and succor us, and support whole legions of plants and animals as they do these things as well.

Indeed, we do not have to do everything ourselves.

There is a vast incredibly rich perennial harvest waiting for us. The solutions for the problems of land degradation, desertification, pollution, drought, and hunger are available to us. They have been here all the time but we in our restlessness have not been able to see them.They are ours for the taking on the investment of a little thought, a little care, and a little patience.

All we have to do is learn how to understand the trees, tap into their promise and potential and use them wisely.

We must also learn not to be greedy with their bounty. There must be room, in our land use patterns and food producing systems, for the wild creatures and wild plants of the world or the world we will bequeath to our children and grandchildren will a poorer, sadder, and more precarious place.

So let us plant trees everywhere.

There is not one ecological problem caused by agriculture or animal husbandry the trees of the world cannot mitigate.

Let each tree planted with thoughtfulness and care, be a step in the greening and healing of the earth.

Amen.

Appendix 1
Environmental and economic potential of Bedouin dryland agriculture

Environmental and economic potential of Bedouin dryland agriculture

A case study in the Northern Negev, Israel

Received 10 September 2007
Revised 12 October 2007
Accepted 7 December 2007

Khalil Abu Rabia
Ben-Gurion University of the Negev, Beer-Sheva, Israel
Elaine Solowey
The Arava Institute for Environmental Studies, Qetura, Israel, and
Stefan Leu
Ben-Gurion University of the Negev, Beer-Sheva, Israel

Abstract

Purpose – The purpose of the paper is to show that land degradation and desertification are threatening the livelihood of more than a billion dryland inhabitants. The paper aims to present traditional and novel approaches for sustainable agricultural exploitation of the arid drylands in Southern Israel and similar climatic zones, and their potential for rehabilitating degraded drylands and increasing agricultural productivity.

Design/methodology/approach – The paper analyses the current agricultural activities on the Abu Rabia farm as well as developing experimental approaches and discusses the expected impact on ecological, economic and social sustainability.

Findings – The farm investigated consists of about 120 hectares of semi-desert land 30 km east of Beer Sheva, divided about 50:50 between rocky hill country and plains with deep loess soil. The area receives an average 200 mm of rain per year. The land is used for raising livestock (about 120 head of sheep and goats), wheat cultivation on high quality soil, and agroforestry, mainly olive cultivation in terraces designed to collect runoff water of seasonal streams. These activities provide a basic income and cover a significant amount of the families' food requirements, but can not provide a full income for a family head in a developed country like Israel. Improving the quality of the grazing land by silvipasture, further investments into high value dryland tree crops and simultaneous production of wood for industry and energy can dramatically increase the farm's income, its resilience to drought and ecological sustainability.

Practical implications – This analysis demonstrates the potential of dryland agroforestry for sustainable development while solving a number of economic and social problems of poor dryland inhabitants, and it contributes to fighting desertification and global warming.

Originality/value – This case study demonstrates that sustainable dryland exploitation by agroforestry can establish significant agricultural production potentials on marginal lands often considered worthless. Because of the establishment of significant and permanent carbon sinks, carbon trading may be mobilized to cover the required investments creating a classical win-win situation.

Keywords Israel, Global warming, Agriculture, Farms, Land, Applied economics

Paper type Research paper

This work was supported in part by a grant from IALC (International Arid Lands Consortium) to E.S. and S.L.

Management of Environmental Quality: An International Journal
Vol. 19 No. 3, 2008
pp. 353-366

1477-7835
DOI 10.1108/14777830810866464

I. Introduction

Land degradation and desertification affect about two-thirds of the world's countries, and 40 per cent of the earth's surface, on which one billion people live (Malagnoux, 2007). The Middle East and the Mediterranean Basin, including Israel and the area analyzed in this paper are among the most severely degraded areas worldwide, with unsustainable, intensive agricultural exploitation beginning 10,000 years ago (Ruddimann, 2003). Most severely affected are the semi-desert areas where biomass cover and soil are dramatically reduced and in extreme cases completely absent. Restoring natural grazing lands as well as establishment of forests, dry woodlands, savannahs or agroforestry projects can dramatically increase the productivity of degraded drylands, even in hyperarid regions that are generally considered unproductive. Such activities also contribute significantly to mitigation of global warming by sequestering carbon into soil and biomass (Lal, 2004; Leu, 1990; 2005).

Techniques for agricultural exploitation of deserts by collecting runoff water in artificially created dams or ditches have been applied for thousands of years in the Arabian Peninsula. Remains of water collection systems many thousands of years old of Israelite, Egyptian and Nabatean origin have been found all over the Israeli Negev Desert (Evenari *et al.*, 1971). Agriculture in the Negev continued during centuries, apparently supported by the Byzantine Empire until the eighth century. The successful restoration of the historical runoff farm at Uvdat (receiving 80 mm of mean annual precipitation) demonstrates the power of these techniques for producing a wide range of agricultural products. It also confirms that the archeological agriculture systems found throughout the Negev desert were designed for current levels of precipitation (Haiman, 1995). In contrast to general beliefs, the Bedouin population in Southern Israel and in the Palestinian Authority has, whenever possible, applied techniques of sustainable desert agriculture throughout the last centuries, and ancient terraces, dams and water holes, many of them recently restored, are testimony to the success and sustainability of those techniques.

Bedouin arrived in Southern Israel hundreds of years ago from Saudi Arabia in search of water and pasture and settled in the largely unoccupied Negev Desert. The land was divided amongst tribal groups according to their influence, power and military strength reflecting their ability to occupy and hold onto sources of water, pastures and the roads leading to the ports and cities (Kressel *et al.*, 1991). Bedouin were well adapted to living in arid environments and were fully aware of the importance of space for pasture and drinking water for man and their livestock. Bedouins adopted agricultural practices by observation of ancient examples and by careful management of the available resources. Agricultural activities and lifestyle were adapted to natural cycles and seasons of the desert. The limitations of scarce water were overcome by capturing and storing water flows in seasonal creeks and rivers during winter, by a system of stow dams and terraces of various sizes. In addition, cisterns were built in rock caves for collection of drinking water. Since maintenance of such structures requires constant supervision, wars, unrest and lack of security were a constant threat to those efforts. Only recently, the traditional dryland agroforestry systems of the Southern Hebron Mountains and the Negev are being restored after hundred years of turmoil.

The current status of biosphere rehabilitation measures and the remaining rehabilitation potential in a 190km^2 section in the Yattir-Lehavim area east of Beer

Sheva (Leu, 2005) was analysed and is documented on this paper as a case study. In this area desertification has been stopped by watershed protection projects, forestry, agroforestry and creation of open woodlands by planting about 100-200 trees per hectare (often termed savannization). The authors have determined and extrapolated the carbon sink potential of equivalent rehabilitation efforts of the semi-desert areas of Israel, the Palestinian Authority and Jordan with similar climatic and topographic conditions and predicted the global carbon sink and biomass production potential of dryland rehabilitation. Conservative estimates yield a potential global sink for 5 gigatons of carbon dioxide per year corresponding to over 30 per cent of the current carbon dioxide accumulation rate in the atmosphere (Leu, 2005). In addition such measures could create sustainable supplies of wood, fodder and food. Thus a global dryland rehabilitation program as demonstrated and proposed here, driven to a large part by investments on private land, could make a significant contribution to mitigating antropogenic CO_2 emissions, stop desertification and give work, food and income to millions of today's poorest populations.

II. Methodology

The location of the Abu Rabbia farm was marked on an Israeli precipitation map (Israel Meteorological Service, Bet Dagan), which indicates a mean annual precipitation at the research site of slightly over 200 mm. The borders of the Abu Rabbia property were marked on a 1:50000 topographical map for identification of the topographical details given in Figure 1. The details on agricultural activities and yields were summarized by the authors. The economic aspects and potential of dryland agroforestry were based on recent studies (Leu, 2005) and on the data gathered during implementation of an IALC (International Arid Lands Consortium) funded silvipasture project on this and similar sites. Data on biological productivity potentials were gathered at public afforestation sites near the research site.

III. Results

Topographical dissection of the Abu Rabbia Farm and implications for agricultural exploitation

The Abu Rabia property encompasses 120 hectares of dryland in the southern foothills of the Hebron Mountains, ranging in elevation between 500 and 600m above sea level. The area receives about 200 mm of mean annual precipitation (the map shown in Figure 2 indicates the average precipitation observed between 1961 and 1991) with the large variations typical for drylands.

Since the farm does not have sufficient water rights for irrigation agriculture, agricultural activities are based on exploitation of the scarce winter rains. About 50ha of relatively flat, deep loess soil are used for cultivation of winter wheat. About 10ha are in direct vicinity of the Abu Rabbia homes and are strongly degraded due to the high density of livestock and poultry. 60ha are rocky, steep hill country used for grazing of the farms livestock. The property borders follow the watersheds feeding the temporary streams of the property (Figure 1), which is an important condition for successful runoff agriculture.

The property encompasses close to 5 km of valleys suitable for construction of terraces and runoff water harvesting. Currently 12 dams have been established, exploiting about 700m (or 15 per cent) of the available potential and creating about 2ha

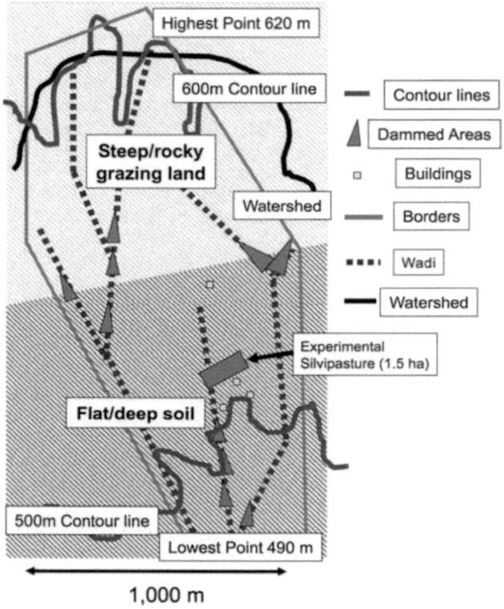

Figure 1.
Schematic overview of the Abu Rabia property with the most significant features (North = up)

Note: The property border in the high country corresponds to the watershed, an essential requirement for full and reliable exploitation of runoff water

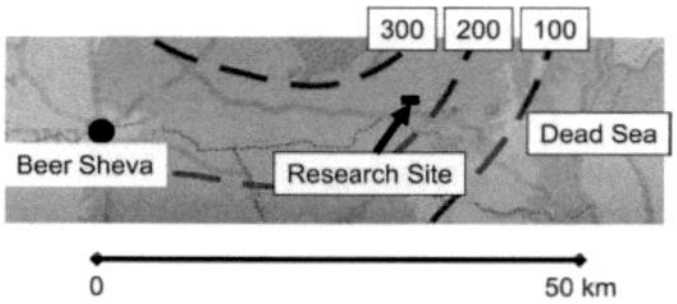

Figure 2.
Geographical location and precipitation level at the research site: the lines of mean annual precipitation (marked in mm/a) are extrapolated from the average precipitation of years 1961-1991

Notes: The averages from 1948-1985 as well as those from 1992-2002 fall well below the average shown here

of well watered land suitable for highly productive agriculture. The remaining valleys as well as part of the rocky slopes provide space for significant increase in runoff agriculture under application of the whole spectrum of harvesting techniques available (Prinz and Wolfer, 1998).

Current agricultural activities

Water management and soil conservation. It takes a commitment to the land and secure land ownership to build the necessary dams or terraces to control water runoff and retain the alluvial silt or soil. Therefore privately held family pieces of land are generally better maintained than equivalent, nearby public lands (leased for example for short term agricultural exploitation), as measured by erosion damage, yields and biodiversity. Private lands are often kept free of erosion, and rocks are removed regularly from the wheat growing areas. Land is plowed along the contours (contour plowing), which allows the rain to soak into the soil and increases soil humidity. Excess water not absorbed by the soil is collected by damming seasonal streams to control the often destructive power of floods. Water slowly absorbed into the soil enhances the natural vegetation that further slows the water flow and enhances soil conservation. A delicate balance is critical in these land management practices, between the amount of water that is captured, and the amounts of water released in a controlled manner. Any flow of rainwater carries topsoil that spreads fertility to dammed areas that become suitable for intensive exploitation. Proper livestock management is crucial in regulating the correct amount of runoff, and organic animal manure accumulating in the grazing lands and in dammed areas further increases soil fertility (Golodets and Boeken, 2006). However, overgrazing especially in dry years can irreversibly damage the land by erosion of topsoil and loss of seeds of native plants and organic litter. Again, based on our survey of the area, private lands are less likely to suffer from overgrazing than public and governmental lands on short- term lease.

Cereal production. The major crop produced on large areas in the Negev is winter wheat. It can yield 1,000-1,500 kg per hectare in average to good years, but the crop has a high risk of failing in years with below average or irregular rainfall. Barley is popular in the cooler and dryer southeast as a substitute for the more valuable, but more drought sensitive wheat. During good seasons the cereals are harvested and surplus is sold, in below average years the grain is not harvested but used as protein rich livestock fodder during summer and fall when the traditional grazing land is exhausted.

Livestock. In order to deal with the harsh conditions of the desert, the Bedouin have specialized in raising animals that can survive with a minimum food and little water, such as camels, sheep or goats.

The Abu Rabia ranch used as a case study in this paper currently holds a herd of about 120 heads of sheep and goats. Even though all of the hill country as well as part of the annual wheat crop are used for grazing, the farm supplies no more than 50-70 per cent of the fodder requirements in average years. Insufficient edible plant material is produced to last the whole year, and by the end of summer the nutritious value of remaining plant material is too low to cover the animals' needs. Additional fodder is bought, but this significantly reduces the profit from the herd.

The most important profits of animal husbandry at the Abu Rabia farm are the production of meat by selling of most of the herd offspring. Dry cheese, for long-term storage, and fresh cheese for immediate use are produced from goat and sheep milk.

Due to the natural, organic fodder and vegetation those animal products are free of chemicals and can be considered organic free range products, a big marketing advantage in the future. Wool is used for production of covers, coats, mattresses and rugs. Women specializing in the wool business can create steady income from the local market that has a significant growth potential.

The total net income of the herd, taking into account labor and fodder costs and veterinary care, is in the range of US$10,000 per year (corresponding to the minimal salary by law in Israel) even though the herd is straining the lands production potential to its limits. Improving the fodder basis of the land appears the only way to increase profitability of the herding business considering the high and growing cost of buying additional fodder.

Dryland agroforestry. The long-term ownership of the land has enabled the Abu Rabbia family to invest in time consuming and expensive techniques of exploiting runoff water for cultivation of fruit trees. So far 12 stow dams have been constructed in dry valleys, creating temporarily flooded areas of totally about 2 hectares (Figure 3).

Most of the family farms grow olives, grapes and figs. Olive oil is used mostly for consumption as food, but also used as healing topical oil and internal medicine and is valued as oil for lighting the home. In addition the terraced areas of the Abu Rabia farm are used for cultivation of a large number of different fruit trees (Figure 4, Table I), whereby olive with about 300 trees is the dominant one.

The gross earnings from the 300 young olive trees is currently about US$2,000 per year (with yields of about 1 l of oil per tree per year), but is expected to reach at least US$10,000 per year with the trees getting bigger and producing 5 l of oil per year and tree. Considering that so far only about 15 per cent of the farm's potential is exploited for agroforestry in this way, olive planting appears to be the most profitable long term option for increasing the farms income in a sustainable way, with gross earnings of over US$50,000 per year anticipated in the long term. However, quality trees cost about US$8 per tree, and investments for terracing, planting, initial watering and fencing

Note: Visible are olive, grape and various fruit trees. On the right side the typical shrub steppe used for grazing livestock

Figure 3.
Dammed and fenced agroforestry plantation at the Abu Rabiah farm with a variety of planted fruit trees

Note: Selection of more robust variants conditions could multiply the yields

Figure 4. Besides the classical dryland species olive (top left) and pomegranate (top right), a wide range of fruit trees manage to produce fruit in such dammed areas, including plums (bottom left), grapes (bottom right), peach, apple, pear, guava, etc. (see Table I)

Species	Produce	Time of harvest	Other benefits
Olive	Oil	November	Leaves and litter for soil improvement and livestock fodder, cutting residues and wood for fires, soil improvement and erosion control
Grapes	Fresh/dry fruit	August	
Almonds	Green fruit	Spring (fresh) or summer (dry	
Figures	Fresh/dry fruit	August/September	
Pomegranate	Fresh fruit	August/September	
Apricots[a]	Fresh fruit	May	
Mulberry[a]	Fresh fruit	June	
Plums[a]	Fresh fruit	June	
Apples[a]	Fresh fruit	June	
Peach[a]	Fresh fruit	May	
Guava[a]	Fresh fruit	September/October	
Pear[a]	Fresh Fruit	May/June	

Note: [a]The species are not generally considered dryland species, but perform well without irrigation

Table I. Fruit tree species grown successfully in the Abu Rabia dams without irrigation (except for a two year adaptation period)

result in total investment costs of US$20 per tree. Since profits are expected earliest 5 years after planting, such investments are a grave burden slowing the development potential for agroforestry.

The terraced areas are used for classical agroforestry activities as well. Vegetables, lentils, tobacco and Sorghum are intercropped with trees in the dammed areas. Due to the large amounts of humidity stored in the dammed areas, tobacco and sorghum continue growing in these dams throughout the summer allowing multiple harvests (Figure 5).

Winter wheat is routinely intercropped in the terraced areas as insurance for drought years, so that even in the driest of years a minimal harvest can be produced on a small area. The yield achieved in those dammed areas is five times that of the regular growing areas and can reach 7.5 tons per hectare.

Expansion and development potential of the agricultural activities

Silvipasture. The grazing potential of the land could easily be doubled or tripled by temporary reduction of grazing intensity and large scale planting of browsing trees and shrubs. Such measures require significant capital for covering the necessary investments and the temporary loss of income. We have initiated a silvipasture project (Figure 6), whereby a number of drought resistant multipurpose trees and shrubs (Table II) are planted mainly on the farm's most degraded land.

The benefits expected from silvipasture are threefold:

(1) The summer active perennials directly supply fresh fodder and protein rich seed pods and litter during the toughest time on the farm, in late summer and fall, when no other sources of protein are available. Sufficient trees can thus substitute for buying fodder and help covering the herd's needs throughout all of the year.

Figure 5.
Classical agroforestry with summer sorghum (left behind) intercropped with olive trees

Note: The humidity stored in the soil of the dammed areas suffices to maintain growth throughout the hot, completely dry summer

Note: Five to tenfold increase in fodder production can be expected on this land after a few years of careful management

Figure 6.
A newly planted mesquite tree (*Prosopis juliflora*) near the farmhouses of the Abu Rabia ranch, established on strongly degraded unproductive grazing land

(2) The trees improve the soil by creating organic litter and improving the mineral and water balance of the land. Synergistic increase of annual vegetation underneath dryland trees is clearly observable, with partial shading, soil improvement and supply of nutrients (nitrogen fixation) all involved in this interesting synergism. The 60ha of hilly grazing lands could easily accommodate some 10,000 fodder trees in a savanna configuration (150 trees per ha), and we expect an increase of at least three-fold in biological productivity on this moderately degraded rangeland. Full exploitation of the silvipasture potential of the Abu Rabia ranch could easily triple biological productivity and thus sustain about double the current herd size. Due to reduced fodder expenses, profitability of the herd could be more than doubled in this way, although an investment of US$20.000-40.000 for planting and temporary reduction of income must be taken into account.

(3) In order to maintain full grazing capacity, the silvipasture trees will have to be trimmed periodically to maintain optimal shading and tree cover. Thus significant amounts of cutting residues become available soon that can be used as fuel. Most rural inhabitants of southern Israel are heating with firewood

Table II.
Some of the silvipasture tree and shrub species planted or considered for planting at the experimental field of the Abu Rabia farm

Species	Properties	Produce	Advantages	Disadvantages
Acacia raddiana	Thorny, highly drought resistant	Edible seed pods and foliage, litter from flowers and leaves	Nitrogen fixing native, biodiversity	Slow growing, long winter dormancy
Prosopis juliflora	Drought resistant tree of American origin	Large edible seed pots and foliage	Nitrogen fixing, fast growing (biomass potential)	Some invasiveness
Pistacia atlantica	Highly drought and cold resistant, almost extinct in the area	Edible seeds, deciduous	Native, biodiversity	
Ceratonia siliqua (Carob)	Evergreen, highly drought resistant	Edible seed pods and foliage	Nitrogen fixing Native	Slow growing
Argania spinosa	Oil producing tree from Morocco, highly drought resistant	Fruit and foliage eaten by livestock	High economic potential of oil production, partly domesticated	Sensitive to *Fusarium*
Pistacia lentiscus	Evergreen small tree/large shrub	Seeds and foliage eaten by livestock	Native, component of the oak/carob/*pistacia*, dry Mediterranean woodlands	Limited drought resistance
Acacia victoria	Australian evergreen	Foliage, seed pods and flowers	Highly drought resistant and cold tolerant, evergreen	Not native slightly invasive
Sclerocarya caffra birrea (Marula)	Fruit and multipurpose tree from Southern African	Fruit, foliage and expensive oil from seeds	Drought and cold tolerant, partly domesticated	
Ziziphus spina-christi	Thorny desert tree with small edible fruit	Fruit and foliage, honey	Extremely drought and heat tolerant	
Acacia bibinosa	Desert shrub	Edible high protein leaves and pods	Drought tolerant, resistant to grazing	
Cassia stuartii	Desert shrub	High protein leaves and pods	Drought tolerant, resistant to grazing	

during the short but cold winter. In times of rising oil prices, increasing wood production for energy is good business with firewood being sold far above the biomass market price of US$100 per ton.

Expansion of agroforestry. The agroforestry potential of the Abu Rabia farm so far has not been realized. Only about 15 per cent of the suitable valleys are watered by the stow dams (Figure 1). The sloping land allows for creation of further terraces or other means to collect runoff water (Prinz and Wolfer, 1998), which could allow plantation of at least 3,000 dryland fruit trees across the farm. Such plantations with a variety of tree species could very dramatically increase the income of the farm. Climate and land are suitable for planting Pistachio, Almonds and Argan, an oil tree from Morocco whose oil is sold at more than US$150 per liter both in Israel and in Europe and the US. Thanks to a first generation of trees with increased yields, and half mechanical oil production equipment available (E. Solowey, personal communication), this tree can represent a high value commodity for Mediterranean dryland farmers.

Biomass production. In an area where heating with wood is the traditional means to pass the cold winter nights, and in the light of increasing oil prices, sale of sustainably harvested wood can further increase the farms income. 15,000 trees planted on the Abu Rabia farm can supply about 100 tons of wood per year from thinning and trimming alone within about ten years. With cut firewood being sold for about US$200 per ton this can represent a very significant additional income. Planting of Eucalyptus or pine (that grow well in the area) for higher yields or saw wood can further increase income from forestry operations without reducing grazing capacity.

Medical plants. Cultivation and exploitation of medical plants has an extremely high economical potential in semi-desert areas. Many aromatic plants of the Oregano family (Figure 7), but also *Foeniculum*, *Artemisia* and others grow well under such harsh conditions without irrigation, and as a fact accumulate more aromatic oil than the same plants grown in wetter environments. The number of such plants can easily be increased by seed application in suitable locations (that are often degraded rocky hillsides), and dried plant material or distilled essential oils can be sold at very attractive prices.

Note: Those plants can easily be established in larger amounts in their natural environment by seed application and protection of the seedlings from excessive competition

Figure 7. Two typical medical plants from the semi-desert environment in Southern Israel

IV. Discussion

Economic growth potential of the Abu Rabia Farm

As demonstrated by the analysis of a private 120ha Bedouin family farm, application of traditional, sustainable dryland agriculture and agroforestry to properties of suitable size and topography in semi-desert environments can improve ecological sustainability, resilience to climate change and multiply the current farm's income. Today's farm income is mainly based on barely sustainable grazing of livestock and provides a single minimum wage of net revenue. Expansion of high value tree crops, increasing the grazing potential by addition of multipurpose trees, commercial sale of wood and further activities like production of organic fruits, eggs or medical plants can create income sufficient for several full time staff even under relatively high salary conditions as those in Israel, with a minimum wage of US$800 and an average gross income of US$1,600 per month.

Significance of dryland farming to the Negev Bedouin

In the Negev alone at least 1,000 similar farms could be created by transferring public agricultural land, that is being exploited under short term lease, to private ownership or long term lease. Such farms could absorb and fully occupy between 3,000 and 5,000 families, without encroaching into ecologically sensitive areas. It should be noted that already now Negev Bedouin successfully exploit lands down to 100 mm of mean annual precipitation using agroforestry techniques as the ones described here. Such a program would be of highest significance to the socioeconomic and cultural development of the Bedouin helping to transform them from a marginalized to a proud, land-owning and productive population. This approach will facilitate integration of women into independent trade and business activities without sudden break of the strict rules of tradition. Orderly transformation of some of the most neglected population segments in the Negev into productive independent farmers based on traditional knowledge and family structures can also help mitigate or avoid some of the social problems observed during the rapid forced urbanization of the last decades (Abu-Saad *et al.*, 2001). By cooperating with like-minded, concerned individuals committed to living on this planet with the least impact on the web of interdependent life, the Bedouin recognize that the human ecology of relationships is no less fragile and in need of a new perspective. They welcome all initiatives that will help them restore natural balance to overexploited drylands.

Global implications

The conclusions presented here concerning the Israeli Bedouin population apply similarly to hundreds of millions of marginalized dryland inhabitants worldwide, who, due to lacking land tenure, conflicts and desertification are driven into vicious cycles of further land degradation, more poverty and thus faster desertification. According to an extensive FAO survey (Pretty *et al.*, 2006), farmers relying on traditional agricultural techniques fared far better than their colleagues relying on intensive, globalized agroindustrial technologies thanks to lower cost (less fertilizer and pesticide use), higher water productivity and higher resilience to drought and weather extremes.

It can be clearly demonstrate here that establishing land tenure, education and instruction in suitable agricultural techniques can create blooming drylands without the need for irrigation, fertilizers and pesticides. Privatization of degrading drylands kept unnecessarily in government hands, under supply and implementation of the necessary rehabilitation technologies, could stop desertification by private initiative alone and create enormous production potentials for food, animal products and carbon negative, sustainable biomass energy (Righelato and Spracklen, 2007). This is in stark contrast to the developments in Israel and many third world countries, where most land is being nationalized at the expense of nomadic people's traditional rights. While other models of land ownership like resource conserving community management (not existing in Israel anymore) could yield similar benefits, privatization has the clear advantage of full accountability concerning all aspects of land management. Only private landowners can, e.g. receive loans for developing their property and subsequently claim carbon credits resulting from land rehabilitation.

Carbon trading for land rehabilitation

The context between land rehabilitation and global warming has been presented sufficiently clear in a large number of publications (Malagnoux, 2007; Lal, 2004; Leu, 1990, 2005). It is clear that suitable land management techniques including tree planting, soil conservation and water management can create very significant carbon sinks in drylands. We estimate that the Abu Rabia property could act as a carbon sink for 200-400 tons of carbon dioxide per year over the next 50 years at least. Accordingly, the investments required for planting and maintenance of the necessary trees, terraces etc. could basically be funded by carbon trading schemes supporting such private efforts. Finding rapid, efficient ways to create certified carbon sinks by rehabilitation of private drylands could probably be one of the safest and most profitable carbon trading practices in terms of verification, guarantees and ecological benefits.

V. Conclusions

Careful land rehabilitation, harvesting of runoff water and investments into suitable tree crops can create viable agricultural production potentials in arid and hyperarid environments, contingent on suitably sized private properties.

Silvipasture can dramatically improve the grazing yield of degraded drylands and results in higher productivity and land rehabilitation and significant carbon sequestration.

Property sizes of 50-100ha per family, and appropriate support for investments and rehabilitation can provide highly climate resilient agricultural production units with significant economic potential.

Land rehabilitation and biomass accumulation on those sites will create permanent carbon sinks that can be sold for covering investment costs.

The findings exemplified here are applicable to all degraded land areas in tropical and subtropical areas and can play a central role in fighting poverty, halting desertification and mitigating global warming.

References

Abu-Saad, K., Weitzmann, S., Abu-Rabiah, Y., Abu-Shared, H. and Fraser, D. (2001), "Rapid lifestyle, diet and health changes among urban Bedouin Arabs of Southern Israel", *Food, Nutrition and Agriculture*, Vol. 28, available at: www.fao.org/DOCREP/003/Y0600M/y0600m06.htm

Evenari, M., Shanan, L. and Tadmor, N. (1971), *The Negev: The Challenge of a Desert*, Harvard University Press, Cambridge, MA.

Golodets, C. and Boeken, B. (2006), "Moderate sheep grazing in semiarid shrubland alters small-scale soil surface structure and patch properties", *Catena*, Vol. 65, pp. 285-91.

Haiman, M. (1995), "Agriculture and Nomad-State relations in the Negev Desert in the Byzantine and early Islamic periods", *Bulletin of the American Schools of Oriental Research*, Vol. 297, pp. 29-53.

Kressel, G.M., Ben-David, J. and Abu Rabia, K. (1991), "Changes in the land usage by the Negev Bedouin since the mid-nineteenth century", *Nomadic Peoples*, Vol. 28, pp. 28-55.

Lal, R. (2004), "Carbon sequestration in dryland ecosystems", *Environmental Management*, Vol. 33, pp. 528-44.

Leu, S. (1990), *Forests and Carbon Dioxide*, Swiss Review of World Affairs No. 2, Verlag NZZ, Zürich, pp. 10-13.

Leu, S. (2005), "Dryland agroforestry for biomass, food, carbon sequestration and desert rehabilitation", *Proceedings of the 14th European Biomass Conference, Paris*, pp. 341-4.

Malagnoux, M. (2007), *Arid Land Forests of the World: Global Environmental Perspectives*, available at: ftp://ftp.fao.org/docrep/fao/010/ah836e/ah836e00.pdf

Pretty, J.N., Noble, A.D., Bossio, D., Dixon, J., Hine, R.E., Penning de Vries, F.W.T. and Morison, J.I.L. (2006), "Resource-conserving agriculture increases yields in developing countries", *Environ. Sci. Technol.*, Vol. 40 No. 4, pp. 1114-9.

Prinz, D. and Wolfer, S. (1998), "Opportunities to ease water scarcity (water conservation techniques and approaches)", *Proceedings, International Conference on World Water Resources at the Beginning of the 21st Century*, available at: www.ubka.uni-karlsruhe.de/cgi-bin/psview?document = /1998/bau-verm/6&search = /1998/bau-verm/6

Righelato, R. and Spracklen, D.V. (2007), "Carbon mitigation by biofuels or by saving and restoring forests?", *Science*, Vol. 317, p. 902.

Ruddimann, W.F. (2003), "The anthropogenic greenhouse effect era began thousands of years ago", *Climatic Change*, Vol. 61, pp. 261-93.

Corresponding author
Khalil Abu Rabia can be contacted at: khalilab@bgu.ac.il

Appendix 2
A Short List of Fuel Trees

Fuelwood species for Arid Areas

Acacia brachystacha
Acacia cambrgi
Acacia cyclops
Acacia nilotica
Acacia radiana
Acacia salinga
Acacia Senegal
Acacia seyal
Acacia tortillis
Adhatoda vasica
Albitzia lebbeck
Anogeissuss latifolia
Azadriachta indica
Cajanus cajun
Cassia siamea
Colophospermum mopane
Emblica officialis
Eucalyptus camaldulensis
Eucalyptus citriodora
Eucalyptus gomphocephala
Eucalyptus microtheca
Eucalyptus occidentalis
Haloxylon aphyllum
Haloxylon persicum
Parkinsonia aculeate
Pinus halepensis
Pithecellobium dulce
Prosopis alba

Prosopis chiliensis
Prosopis cineraria
Prosopis juliflora
Prosopis pallida
Prosopis tamarugo
Tamarix aphylla
Tamarix nilotica
Zisiphus maurinia
Zisiphus spina-christi

Fuelwood species for Tropical Highlands

Acacia mearnsii
Ailanthus altissima
Alnus acuminata
Alnus nepalensis
Alnus rubus
Eucalyptus globules
Eucalyptus grandid
Grevillia robusta
Inga vera
Shorea spp.

Fuelwood species for Humid Tropics

Acacia auriculiformis
Calliandra calothyrus
Casuarina equistefolia
Derris indica
Eugenia jambolana
Gliricidia sepium
Gmelina arborea
Guazuma ulmifolia
Leucaena leucocephalia
Mangroves

Mimosa scabrella
Muntingia calabura
Sesbania bispinosa
Sesbania grandiflora
Syzygium cumini
Terminalia cataapa
Trema spp.

Glossary

Actinobacteria—nitrogen fixing bacteria related to fungi
Actinomycete-a specialized nitrogen fixing micro organism related to fungi
Abcisse—the separation and fall of fruit, flowers, leaves and other plant parts
Agricultural pyramid—the sides of the agricultural pyramid are: plants, farmers, animals, soil
Allele—different versions of the same gene, sometimes in the same organism
Aquaculture systems—aquatic cropping systems to produce plants, mollusks or fish
Arable soil—soil that is suitable for agriculture
Artificial manure—the original name for chemical fertilizer
Attush-bark cloth garment from northern Japan
Bast-the inner bark of certain trees
Bioinnoculants—cultures of bacteria and soil symbionts added to compost
Biometric School—the school of plant genetics dedicated to population breeding
Biosphere—the earth and all living elements upon it
Biotechnology—-the science of accelerated, trans-genetic manipulation of living creatures
Breeding—Controlled propagation of palnts to achieve specific purposes
Cash crops—crops which are cultivated to supply raw materials in return for cash payment
Cellulose digesters—modern closed compost bins

Cellulosic ethanol-bio fuel derived from agricultural wastes and woody sources
Cloches—frames covered with glass or plastic to raise the air temperature around plants
Cold bed agriculture—a system in which plants are nourished with cold water
Companion planting—the planting of different cultivars which benefit from each other's proximity
Compost—broken down, rotted plant material suitable for inclusion in the farm nutrient cycle
Conventional agriculture—-agriculture based on the use of fossil fuels, fertilizer in the form chemical salts and monocrops
Conventional breeding processes—pedigreed breeding resulting in huge stands of identical plants
Crop diversity—multiple cultivars on a farm, multiple varieties within a cultivar
Crop rotation—planting crops in planned succession to curtail insect activity and preserve fertility
Cultivars—cultivated, domesticated plants
Cultivated—raised purposefully by the farmer or agriculturalist, worked by the farmer
Cyanobacteria—blue green algae
Deforestation-the removal of forest cover and vegetation
Deciduous—plants that lose leaves in the winter
Diazotrophs-microorganisms that can "fix" nitrogen
Diazovesicles-specialized plant organs for absorbing nitrogen made available by bacteria

Domestication—the process by which the breeding and propagation of a wild organism comes under human influence
Dormancy—a state of inactivity or suspension of biological processes
Drip irrigation—irrigation systems which deliver precise amounts of water through small pipes
Ecotypical cultivar—a strain of crop plants well adapted to a specific locale
Epistasis—one gene modifies the expression of another gene, not an alelle of the first
Erosion- the loss of topsoil to wind and water
Ethanol—ethyl alcohol, C2H5OH
Field resistance—enduring, non-specific, polygenetic tolerance of disease or environmental challenge
Frankia-a type of nitrogen fixing bacteria associated with alder trees
Gene transfers—the transfer of genetic material to a different organism's genome
Genetic base—the gene pool of a crop plant
Genetic diversity—the variability and flexibility of a crop's gene pool
Genetic engineering—genetic modification, GM
Genetic modification—manipulation of the genome of an organism by the addition of exogenes
Genome—the complete complement of genes within an organism
Genotypes—two organisms with the same genome are said to have the same genotype
Germinate—to sprout or develop
Germplasm—germ cells and bearers of heredity

Gluten—elastic protein substance in wheat that gives cohesiveness to dough
Gray water—urban but non-industrial waste water, mostly from private dwellings
Growth inhibitors—substances which keep a seed from sprouting, or a plant from growing
Halophytes—salt loving or salt tolerant plants
Harvest index—-the proportion of a crop which can be used for the crop's principle purpose
Heirloom variety—older variations of cultivated plants with specific characteristics that breed true
Heliculture-the cultivation of snails
Hybrids—the progeny of a cross between two different varieties
Hydrolysis- molecule is split because of the addition of water
Hydroponics—literally 'water working', growing plants in water
Hyphae-(plural) fungal filaments
Indigenous—native to an area
Industrial style farming—high input, high energy use factory style farming
Integrated crops—crops and crop successions which are planted together for synergistic benefits
Integrated pest management—limiting pest populations by good farm management,
Intercrop—growing two or more cultivars together for their mutual benefit
Isolines—static and uniform end result of pedigreed breeding

Jackpot traits—the 'jackpot' traits are: drought tolerance, salt tolerant and nitrogen fixing ability
Jaggery—a form of palm sugar
Landraces—primitive ecotypical cultivars, with great genetic variability and diversity
Larva- the juvenile stage of a metamophizing insect, larva pupate then transform into adults
Legumes—a plant in the pea or bean family, usually capable of fixing nitrogen
Lichen- organisms composed of algae and fungus in symbiotic self replicating stable states
Lignin-complex chemical compound mostly found in wood
Masi-a form of Polynesian bark cloth
Masaka-African bark cloth
Mendelian School—the pedigreed school of plant breeding
Microgardening—producing food in tiny private areas
Modified seeds—seed which has undergone genetic manipulation
Monocrops—crops which are not genetically diverse, frequently the only cultivar in a wide area
Monoculture—the practice of growing large stands of one kind of identical plant
Mycellium—the vegetative part of fungi, made up of hyphae
Mycorrhiza—fungal soil symbiont microorganisms
Nymph-the juvenile, miniature form of some insects, that do not pupate, but grow into larger

forms by shedding their skin until they reach adulthood

Organically grown crops—crops which are raised in healthy soil without chemicals of any kind

Parasitize—to become a parasite upon another living organism

Patented genes—genes which are 'owned' by companies, individuals or institutions

Pattern rotation—a sophisticated form of crop rotation which curtails pests and disease

Perennial—a plant which lives for many years

Permaculture—the art/science of creating a stable, sustainable, humane, energy efficient food web

Phyto-pesticides—target materials derived from plants i.e. Neem

Plant clones—identical plants produced asexually

Pleiotrophy—effect of a gene on the expression of a number of different genetic traits

Polyunsaturated—oil or fatty acid rich in unsaturated bonds

Propagation—the breeding and multiplication of living organisms

Reclamative crop—a crop which is planted to improve impoverished soil

Refugia-areas of surviving population of rare species

Relic crop—a rare ecotypical cultivar, a surviving 'lost crop'.

Relict species-species left unchanged and isolated by changes in climate or by geography

Resistant varieties—plants which demonstrate non-specific or field resistance
Rhizobia-nitrogen fixing bacteria associated with legumes
Rhizosphere-the root zone of some plants
Riparian systems—relating to or involving a watershed
Saponins—any of various glucosides found in plants
Seed Bank—an institution dedicated to the preservation and storage of seeds
Sericulture-cultivation of silkworms
Shelf-life—the amount of time a food or product remains edible before spoilage
Silwan-date syrup
Siapo- highly developed form of bark cloth
Soil building crops—crops which improve the health and fertility of the soil
Species—a class of individual organisms having common attributes and designated by the same name
Stem flow-water which is collected by plants from moist air by condensation on the physical structure of the plants
Subsistence agriculture—agriculture in which families and villages provide food for themselves
Substantial equivalence—the idea that food plants are not significantly changed by genetic manipulation so do not have to be tested or investigated vis a vis human health
Sustainable agriculture—agriculture which does not damage soil, water or farm environment

Synergistic integration—farm elements which enhance the health and productivity of other farm elements
Tertiary foods—herbs, spices, tea, coffee, cola and chocolate
Toddy-a drink made from fermented palm sap
Toxicity—a state relating to or caused by a toxic substance
Toxin—a poisonous substance which is the product of the metabolic activity of a living organism
Transgene—gene from a dissimilar organism
Transgene traits—traits acquired by transferring genes from another organism
Truffles-edible fungi that grow on the roots of trees and other perennial plants
Truffieres-truffle producing orchards
Tubers—thickened underground stems of a plant
Vermiculture-the cultivation of earthworms or red worms
Vertisols—'black cotton' clay soils which are difficult to make productive
Wildcrafting-taking a harvest from wild plants
Wildcrop—plant product which is not taken from cultivated plants but gathered from the wild
Wild pathosystems—in which each plant is different so the spread of disease is curtailed by variability
Windbreak—a line of trees or vegetation planted to reduce the force of the wind
Xiji-a tenured wildcrafting territory for collecting frankincense gum

Yield drag—the harvest deficit between unmodified crops and GM crops due to GM plants' resource debts

Bibliography

Introduction

Diamond, Jared, *Collapse, How Societies Choose to Fail Or Succeed*, Viking Publishers, 2005

Kunstler, James, *The Long Emergency*, Atlantic Monthy Press, 2005

Smith, J. Russel, *Tree Crops, A Permanent Agriculture*, Harper Colophon Books, 1950

Chapter 1

Giono, Jean; Giono, Aline, *The Man Who Planted Trees*, Shambhala Publication Inc, reprint, 2000

Glueck, Nelson, *Rivers In the Desert: A History of the Negev*, Norton Publishers, 1968

Kitterage, Joseph, *Forest Influences, The Effects of Woody Vegetation on Climate Soil and Water*, Dover Publication Inc., reprint, 1973

Oryx, The International Journal of Conservation, April 2002, Volume 36 Number Two "Rainforests Harvest the Skies" p. 110

Chapter 2

Jacke, David, *Edible Forest Gardens*, Chelsea Green Publications, 2008

Leu, Stefan, *Forests and Carbon Dioxide*, Ben Gurion University Press, 2005

Prinz, Dieter, *Ron-off Farming*, WCA infoNET, 2005

Chapter 3

Ajayi OC, Place F, Kwesiga F and Mafongoya P, 2006, Fertilizer Tree Fallows in Zambia, Occasional Paper No. 5, Nairobi, World Agroforestry Cnter

ICRAF, *Improving on Improvement, Mixed Tree Fallows for better Maize Crops in South Africa*,2006, The International Centre for Research In Agroforestry

Jacke, David, *Edible Forest Gardens*, Chelsea GreenPublications, 2008

National Academy of Sciences, *Tropical Legumes: Resources for the Future*, National Academy Press, 1979 Washington D.C
Raven, Peter and Johnson, George, Biology, Times, Mirrot/Mosby College Publishing 1989
Smith, J. Russel, *Tree Crops, A Permanent Agriculture*, Harper Colophon Books, 1950

Chapter 4

Biology of the Acacia, Australian Systematic Botany Vol. 16, No. 1, 2003
Harwood, Rinaudo, Adewusi, *Developing Australian Acacia Seeds as Human food for the Sahel*, CSIRO, Department of Forestry, Canberra, 2007
House, APN and Harwood, ED (eds), *Australian Dry Zone Acacias for Human Food*, CSIRO, Division of Forestry, Canberra
Logan, William Bryant, *Oak, the Frame of Civilization*, W.W.Norton and Company, 2005

Chapter 5

Blume, David, with Michael Winks, and R. Buckminster (FWS) Fuller, *Alcohol Can Be a Gas!: Fueling an Ethanol Revolution for the 21st Century*, The International Institute for Ecological Agriculture
Crooks, Anthony C., Cooperatives and New Uses for Agricultural Products: An Assessment of the Fuel Ethanol Industry, USDA (Kindle Edition - Mar. 17, 2010) – Kindle
Eckholm, EP. *Planting for the Future, Forestry for Human Needs*, Worldwatch Paper 26, 1979, Worldwatch Institute, New York, NY
Goettemoeller, Jeffrey and Adrian, *Sustainable Ethanol: Biofuels, Biorefineries, Cellulosic Biomass, Flex-fuel Vehicles, and Sustainable Farming for Energy Independence*, 2007Prairie Oak Publishing
National Academy of Sciences, *Lost Crops of Africa*, Vol. 1 Grains, National Academy Press, 1996

Chapter 6

Fleuret, P. and Fluert, A., *Fuelwood Use In a Peasant Community*, Journal of Developing Areas 12(3): pp. 315-322

National Academy of Sciences, *Lost Crops of Africa*, Vol. 1 Grains, National Academy Press, 1996

National Academy of Sciences, *Firewood Crops Shrub and Tree Species for Energy Production*, National Academy Press, 1980

Chapter 7

Karg, Sabrine, "Direct Evidence of Heath Land Management in the Early Bronze Age: From the Grave Mound Skelhoj in Western Denmark," *Vegetation, History and Archeology*, Vol. 17 No. 1, January 2008, Springer Publishers

Soule, Judy, Piper, Jon and Jackson, Wes, *Farming in Nature's Image: An Ecological Approach to Agriculture*, Island Press, (Paperback - Dec. 1, 1991)

Chapter 8

O'Lenick, Anthony, et. al., *Oils of Nature*, Allured Publishing, 2008

Train, John; Train, Maria Teresa, *The Olive, Tree of Civilization* (MTT SCALA); illustrated edition (October 30, 2004)

Chapter 9

"Bark cloth," *The Columbia Encyclopedia*, Sixth Edition. 2008, Encyclopedia.com. (March 21, 2010), http://www.encyclopedia.com/doc/1E1-barkclot.html

Chapter 10

Amundsen, Roald, *Race to the South Pole (The Great Adventures)* White Star; illustrated edition (March 13, 2007)

Carpenter, Kenneth J., *The History of Scurvy and Vitamin C*, Cambridge University Press (April 29, 1988)

Frankenburg, Frances R. M.D., *Vitamin Discoveries and Disasters: History, Science, and Controversies*, The Praeger Series on Contemporary Health and Living, Praeger; 1 edition (July 23, 2009)

Chapter 11

Stearn, William T., *Stearn's Dictionary of Plant Names for Gardeners: A Handbook on the Origin and Meaning of the Botanical Names of Some Cultivated Plants*, Timber Press Limited, November 1, 2002

Chapter 12

Burbank, Luther, *Trees Whose Products are Useful Substances: From the Sugar Maple to the Turpentine Tree*, Athena University Press (July 15, 2004)

Dyer, E. H., *Sugar: Sugar, Sucrose, Molasses, Sugar Beet, Glycomics, Rum, History of Sugar, Jaggery, Bagasse, Brown Sugar, Palm Sugar, Sugarloaf* [Paperback], Books LLC (May 22, 2010)

Chapter 13

Marcu, Monica G., *Miracle Tree*, KOS Health Publications (May 2005)

Pirie, N. W., *Leaf Protein and Its By-products in Human and Animal Nutrition*, Cambridge University Press; 2nd Edition (January 21, 2008)

Chapter 14

Coe, Sophe and Michael, *The True History of Chocolate*, 2nd Edition, Thames and Hudson, 1993

Groom, Nigel, *Frankincesne and Myrrh*, Longman, London and New York, 1981

Rocco, Fiametta, *The Miraculous Fever-Tree: Malaria and the Quest for a Cure That Changed the World*, Harper Collins, 2003

Weinberg, Bennett Alan; Bealer, Bonnie K., *The World of Caffeine: The Science and Culture of the World's Most Popular Drug*
Zohar, Amar, *The Book of Incense* (Hebrew), Oren Hapcott Publishers, Tel Aviv, 2004
Zohar, Amar, *Balm for the Body* (Hebrew), Oren Hapcott Publishers, Tel Aviv 2004
Taylor, Norman, *Plant Drugs that Changed the World*, Dodd, Mead and Company, 1965

Chapter 16

Ditlhogo, M., Allotey, J., Mpuchane, S., Teferra, G., Gashe, B.A. and Siame, B.A. (1996), “Interactions between the mopane caterpillar, *Imbrasia belina*, and its host, Colophospermum mopane in Botswana,” *In Flower*, C., Wardell-Johnson, G. and Jamieson, A. (eds.), Management of mopane in southern Africa. Ch.9, 46-49
Latham, Paul, from the English edition of his manual *Edible caterpillars and their food plants in Bas Congo* (1999
Reddy, R.M., 2010. *Conservation need of tropical tasar silk insect, Antheraea Mylitta drury (Lepidoptera: Saturniidae)-strategies and impact.* J. Entomol., 7: 152-159, DOI: 10.3923/je.2010.152.159 URL: http://scialert.net/abstract/?doi=je.2010.152.159
www.food-insects.com
“Silkworm,” *The Columbia Encyclopedia*, 6th Edition 2008. Encyclopedia.com. 11 Mar. 2010 <http://www.encyclopedia.com>.

Index

A

B

C

D

E

F

G

H

I

J

K

L

M

N

O

P

Q

R

S

T

U

V

W

X

Y

Z

Made in United States
Troutdale, OR
05/27/2025

31705344R00226